FOREST, GOVERNMENT AND TRIBE

About the Editor

Dr. Chittaranjan Kumar Paty, an alumnus of Ranchi University, has been teaching History since 1970. He is presently the Reader and Head of the Department of History, Tata College, Chaibasa. He was awarded the Ph.D. degree in 1991 by Ranchi University. He has published a book entitled *History of Seraikella and Kharsawan States* (2002). He has contributed several research papers to different journals and books. Presently, he is working on the tribal history of Jharkhand.

FOREST GOVERNMENT AND TRIBE

Edited by
Dr. Chittaranjan Kumar Paty
Reader & Head, Dept. of History,
Tata College, Chaibasa.

CONCEPT PUBLISHING COMPANY, NEW DELHI-110059

Cataloging in Publication Data—DK
Courtesy: D.K. Agencies (P) Ltd. <docinfo@dkagencies.com>

Forest, government and tribe / edited by Chittaranjan Kumar Paty.
p. cm.
Papers presented at a national conference held at Chaibasa during 2-3 March 2005.
Includes bibliographical references.
Includes index.
ISBN 8180694062

1. Community forestry—India—Congresses. 2. Community forestry—India—Jharkhand—Congresses.3. India—Scheduled tribes—Legal status, laws, etc. 4. Jharkhand (India)—Scheduled tribes—Legal status, laws, etc. 5. Forest policy—India—Congresses. 6. Forest policy—India—Jharkhand—Congresses. 7. Human ecology—India—Congresses. 8. Human ecology—India—Jharkhand—Congresses. I. Paty, Chittaranjan Kumar, 1946-

DDC 333.750 954 127 22

ISBN-81-8069-406-2

First Published 2007

Published and Printed by
Ashok Kumar Mittal
Concept Publishing Company
A/15-16, Commercial Block, Mohan Garden
NEW DELHI-110059 (India)
Phones: 25351460, 25351794
Fax: 0091-11-25357103
E-mail : publishing@conceptpub.com

Contents

INTRODUCTION

The papers in this book have been contributed to the National Conference organised by the Department of History, Tata College, Chaibasa on 2-3 March, 2005 and sponsored by the UGC, Eastern Regional Office, Kolkata. The theme of the Conference was Forest, Government and Tribes.

As Chaibasa is surrounded by thick jungles and the Singhbhum region is predominantly inhabited by tribal people I decided to work on them. At the same time all the tribal movements which had occurred in the area hinged around the forest. The whole life of the indigenous people depended upon these forests. Thus the central theme of the conference took shape.

The theme was divided into four subparts: Forest Laws, Ecological Ideologies, Social Protest and Forest and Development. The conference emphasised ideas and issues relevant for a closer understanding of the forest, government policies and the forest dwellers. In this conference we welcomed the different papers of learned scholars. So the collective wisdom of the participants is also expected to provide guidelines to the administrators and policy-makers to prepare future policies related to jungles for the development of the indigenous people.

The discussion of this monograph begins with the paper of Prof. Buddhadeb Choudhury which highlights the forest laws of the country in the context of exploitation of forest resources on which the tribal economy is based. He rightly observes that the National Forest Policy should protect the autonomy right of the tribal people to use forest resources.

Sri S. Bosu Mullick concentrates on the forest policy of Jharkhand which not only changed the existing life of the people but also adversely affected the whole eco-system of the region. Due to these changes, he affirms the tribals of this region were reduced to ecological refugees causing social protest ultimately for leading to the self-rule in Jharkhand.

Sri A. K. Sinha studies forest laws in historical context as causing the degradation of the forest. He suggests that only such laws should be enforced which are ecology-friendly.

Dr. P. P. Mahto and Dr. K. C. Mahto describe that the tribals of the Eastern India especially Jharkhand are inseparably linked to the forest around them. Their totems, folk songs and myths are inspired by forest. But, after independence, government-sponsored multi-purpose projects destructed the ecology and bio-diversity of the people.

After defining ecology in the socio-philosophical context, Dr. P. Sen explains that the long association with forest and nature induced adivasis to evolve an eco-centric belief system. The paper finally enumerates their forest-related ecological sensibility and an implied ecological morality towards the nature as a whole.

Sri Amit S. Paty takes up the Lodha tribe of West Bengal to highlight the sad reality of the degeneration of a forest dependent and peace-loving tribe into a criminal tribe when they were systematically dislodged from their natural setting.

Dr. A. Mahanta explains the ecological ideologies in tribal folk songs. He observes that the folk songs not only draw on forest but they also relate tribal methods of treatment of various diseases.

In the back-drop of conflicting ecological ideologies of the government and Ho tribe Dr. A. K. Sen takes up select forest laws as contributing to social protest. He underlines how the legalistic popular assertion of customary forest rights emphasised the very notion of Ho cultural identity.

Dr. C. K. Paty traces the close link of the economy, social and culture of the tribals of Singhbhum with the forest. But the recent encroachments upon their Khuntkatti rights, exploitation of their women, displacements and the plantation of teak trees led to their collective protest which not only turned out to be anti-government but in a way destruction to the forest also.

Ms. Chumki Piplai highlights the forest dependent hunter gatherer Jarawa people of Andaman. She argues that this harmonious relation underlines the reason why they never destroy the forest as they consider it to be the very source of their existence.

Sri S. K. Bhattacharaya's field-work is on the migrant Munda tribe of Baratang island of Andaman called the Ranchiwallas. These people settled in a remote area and converted a barren land into partly cultivable and forested tract.

Sri A. K. Verma describes the brief history of the forest of Chhotanagpur. He explains the socio-economic role of forest in the life of tribal people of this region.

Dr. A. Shrivastava highlights the symbiotic link between forest and tribe manifested in their survivalistic appropriation of MFP. She maintains that profit-making tendency of non-tribal traders and government agency has systematically disrupted this long basis of tribal material life.

Sri Dipak Piplai explains that forest is not only the means of revenue earning but truly a protector of human life also. All the social, economic and cultural practices closely revolve round the forest but, due to deforestation by the urban high-ups, the forest is fast becoming barren thereby throwing a serious challenge to the society at large.

Emphasising the social-economic and medicinal use of different trees, fruits, barks and related things of the forest, Sri D. K. Mishra explains the reasons behind the systematic degradation of the forest and ends the discussion by suggesting the ways how the forest should be conserved.

In this way, these papers present a comprehensive picture of forest, government and tribe. They provide clear guidelines for bringing about all-round development of these people in the coming time.

Date: 3 March 2005

CHITTARANJAN KUMAR PATY
Department of History
Tata College, Chaibasa.

ACKNOWLEDGEMENTS

This anthology owes much to several persons and sources. First of all, I am thankful to all the contributors, whose research papers have made it possible to present this monograph. I am grateful to the University Grants Commission, Eastern Regional Office, Kolkata for their financial assistance without which the conference and the publication could not have taken shape . I am obliged to Dr. G. P. Rajwar, the Principal of Tata College, for his active help in this regard. I am extremely indebted to Dr. A. K. Sen and Dr. Padmaja Sen for their valuable advice and guidance which encouraged me to bring the work to a successful close. I owe my thanks to Prof. A. Dubey for his linguistic help and keen interest which he has always shown in my work. I must thank my friend Prof. Aditya Kumar Mishra, Principal, Bahragora College, for his suggestions and encouragements. I extend my sincere thanks to Dr S. P. Singh, Prof. R. L. Agarwal, Prof. K. Boipai, Dr. Aruna Shrivastava and Dr. Lok Nath who helped me in organising the conference. Thanks are also due to Prof. Ahmad Raza, Dr. M. A. Khan, Dr. Tapan Khanrah, Dr. S.C. Dash, Sri Vijay Karmakar, Sri Gopi Singh, Sri Manohar Hembrom and Ms. J. Soy who helped me from time to time. I am grateful to Sri Ajay Prasad of CMC Computer Education, Chaibasa center and Sri Viswanath Dodrajka of Vivek Bharti book shop, Chaibasa for the preparation of the monograph. At the end I express my thanks to all the members of my family for their active support towards completion of this project.

The Contributors

Adikanta Mahanta is a Researcher from Chitrada, Mayurbhanj, Orissa.

Amit Saurabh Paty is M.A. in Anthropology and pursuing research on Tribal Life.
E-mail: Amit_paty@rediffmail.com

Aruna Srivastava is Reader and Head, Dept of Anthropology, Tata College, Chaibasa, Ranchi University, Ranchi.

Ashok Kumar Sinha is Reader and Head, Dept of History, Co-operative College, Jamshedpur, Ranchi University, Ranchi.

Ashok Kumar Verma is Reader and Head, Dept of History, L.B.S.M. College, Jamshedpur, Ranchi University, Ranchi.

Asoka Kumar Sen former Professor & Head, Dept of History, Tata College, Chaibasa and now an independent Researcher of Tribal History.
E-mail: asokakumarsen@sify.com

Buddhadeb Chaudhury is Ambedkar Chair Professor, Dept of Anthropology, Calcutta University, Kolkata.
E- mail : chau@cal2.vsnl.net.in

Chittaranjan Kumar Paty is Reader and Head, Dept of History, Tata College, Chaibasa, Ranchi University, Ranchi.
E-mail : patyck@rediffmail.com

Chumki Piplai is a Research Associate, Anthropological Survey of India, Kolkata.

Dinesh Kumar Mishra is Reader, Dept of Economics, Co-operative College, Jamshedpur, Ranchi University, Ranchi.

Dipak Piplai is a Researcher and Social Activist, Kolkata.
E-mail : piplaidipak@hotmail.com

Kshirod Chandra Mahato is a Reader, Dept of Bengali, Subernarekha Mahavidyalaya, Gopiballavpur, West Midnapur, West Bengal.

Padmaja Sen is Reader and Head, Dept of Philosophy, Mahila College, Chaibasa, Ranchi University, Ranchi.
E-mail : padmajasen@sify.com

Pashupati Prasad Mahato is Chairman, Anthropological Research Committee, Indian Academy of Social Sciences, Kolkata.

Sanjay Bosu Mullick is Senior Asst Director, William Carey Study & Research Center, Kolkata.
E-mail: rch_sanjay@sancharnet.in

Sapan Kumar Bhattacharya is an Assistant Anthropologist, Anthropological Survey of India, Kolkata.

1

FOREST AND TRIBALS
A Historical Review of Forest Policy

BUDDHADEB CHAUDHURI

In the Indian subcontinent the term forest does not merely mean an area covered with trees but it carries the impression of an entity that is a sum total of ecological and biological parameters. According to the cultural traditions of the country, forests hold a very important place in the life and philosophy of the people apart from the material gains that it brings about continuously.

The forests in India according to one estimate, cover an area of 7,43,584 sq. kms. or 22 per cent of the total geographical area of India. Among states, only Tripura has 60.8 per cent and Orissa 43.74 per cent of the total area under forests. In all other states, it is less than 40 per cent. Again, many of the forests are in poor condition and according to some recent official estimates, the forests do not cover more than 13 per cent of the total area of India.

In the recent past, forest has become an issue of common concern in India mainly for three reasons:

(1) Due to ecological consideration.
(2) For its commercial value.
(3) In view of the growing unrest over centring forest and exploitation of forest resources.

The controversial Forest Bills of 1980/1988 have also generated enough criticism as well as concern and awareness among all.

Due to inexorable forces of history, groups of people of some ethnic origin were pushed to dense forest, hilly, inaccessible and inhospitable terrain. Environmental constraints made such groups insular and kept them cut off from the main stream of the people resulting in their under-development. The environment influenced culture and life style of such people and played a great role in providing food and shelter to them. These facts were discernible and were recognized and such people were grouped as Tribals, Aboriginals, Girijan, Vanabasi etc.

The Governments being conscious of the under-development of the tribals, took special measures to study their problems and initiated executive and legislative action to safeguard their interests and to create conditions for their development culminating in providing even constitutional safeguards.

Articles 15(4), 17, 19(5), 46, 164(1), 244, 244A, 275, 20, 332, 334, 335, 338, 339, 342, 344, 371A, 371B and 371C of the Constitution of India make specific references to the Scheduled Castes and Scheduled Tribes.

Articles 15(4), 17 and 16 make provisions for promotion of educational and economic interests and of protection from social injustice and all forms of exploitation.

Article 339(2) prescribes that the executive power of the Union shall extend to the giving directions to a State as to the drawing up and execution of schemes specified in the direction to be essential for the welfare of the Scheduled Tribes in the State.

Central and State policies, so far as forests and tribals are concerned emanate from the above constitutional obligation as a welfare state.

The thrust, therefore, in the 5-year plans has been to end exploitation of the tribals and to initiate schemes so that they may go up the poverty line. State Governments on the directives of the Central Government and on its own, initiate several schemes to provide gainful employment to the tribal, to open up lines of communication, provide health facilities, educational facilities etc. and to take economic measures for development.

In keeping with the above strategy for development of the tribals some of the forestry schemes of the Tribal Sub-plan areas are as follows:

(1) Plantation of quick growing species.
(2) Social Forestry including Rural Firewood plantation.
(3) Special SIDA aided Social Forestry Project.
(4) Roadside Plantations including urban forestry.
(5) Development of Minor Forest Produce.
(6) Wild – life and environmental preservation.
(7) Soil Conservation.
(8) Lac Development.
(9) Tassar Development.

With a view to end exploitation of the tribals in the context of marketing of forest produce several steps have been taken by the Forest Department. Some of the salient features are as follows:

(1) Abolition of Contract System in forestry operations.
(2) Collection and marketing of Minor Forest Produce.

Rights of the tribes were inquired into and recorded. Over and above, tribals were permitted free grazing, collection of flowers, fruits, roots and tubers and removal of firewood for their domestic consumption before the introduction of amendment regarding Non-Forest Life in Forest Conservation (Amendment) Act, 1990. This is in short the official version of tribal and forest situation; but the reality is some where else. Therefore a resume of forest management since British days will not be out of place here to appreciate environment – forest – tribal nexus.

Basic Objectives

The basic objectives that should govern the National Forest Policy are the followings:

Maintenance of environmental stability through preservation and where necessary, restoration of the ecological balance that has been adversely disturbed by serious depletion of the forests of the country.

Conserving the natural heritage of the country by preserving the remaining natural forests with the vast variety of flora and fauna, which represent the remarkable biological diversity and genetic resources of the country.

Checking soil erosion and denudation in the catchments areas of rivers, lakes, reservoirs in the interest of soil and water conservation, for mitigating floods and droughts and for the retardation of siltation of reservoirs.

Checking the extension of sand - dunes in the desert areas of Rajasthan and along the coastal tracts.

Increasing substantially the forest/tree cover in the country through massive afforestation and social forestry programmes, especially on all denuded, degraded and unproductive lands.

Meeting the requirements of fuel wood, fodder minor forest produce and small timber of the rural and tribal populations.

Increasing the productivity of forests to meet essential national needs.

Encouraging efficient utilization of forest produce and maximising substitution of wood.

Creating massive people's movement with the involvement of women for achieving these objectives and to minimize pressure on existing forest areas.

The principal aim of forest policy must be to ensure stability and maintenance of ecological balance including atmospheric equilibrium which are vital for sustenance of all life forms, human, animal and plant. The derivation of direct economic benefits must be subordinated to this principal aim.

Dhebar Commission Recommendations

The different Five Year Plans gave a new boost to the forest management and the increased demand of timber and various forest products on one hand and pressure on land on other resulted in substantial loss of forest land and forced the administration to embark on new policy again. In 1961, the Dhebar Commission had already recommended that subject to safeguards, tribals should be allowed forest lands for cultivation, their needs should be met from outlying areas in the reserve forests and their requirements for grazing and shifting cultivation should be conceded. The Commission wanted the forest department to participate in the betterment of the tribals side by side with the development and conservation of the forest. Their intention was to provide tribals with work all the year around by the Forest Department in consultation with Agriculture, Industries and the Development Departments in each region.

Shiloo Ao Committee in 1969 more or less reiterated the recommendation of the Dhebar Commission and deplored the inaction and apathy of the administration in improving the lot of the tribal people.

National Agriculture Commission

The National Commission on Agriculture in 1976 accepted the inter-relationship of forest economy with rural and tribal economy, but it also accepted that Forestry needed strengthening by rationality in operations, larger investments and use of newer techniques so that forest can yield a higher surplus to be shared locally, regionally and nationally as against rapid disappearance of trees and vegetation. The Commission recommended revision of the 1952 National Forest Policy by incorporating significant shifts and stresses in the forest policy arising out of the recommendations of the Central Board of Forestry. The Estimate Committee (1968-69) of the Fourth Lok Sabha in its 76th Report also recommended that the National Forest Policy should be reappraised so as to make the new policy more purposeful, realistic, effective and operative for the development of forests and forestry in the country.

42nd Amendment in Constitution

During 1975-76 National Emergency, the 42nd Amendment of the Indian Constitution was made by inserting Article 48A in the part IV of the Indian Constitution. This amendment suggests the state to endeavour to protect and improve the environment and safeguard the Forests and wild life of the country. It has direct bearing on the life of Scheduled Tribes because they live mostly in forest areas and for them this Amendment is almost silent. This led to large scale agitation in tribal areas.

A Resume of Forest Policy

Tribal and forest have become the two inseparable words. Vanbasi, Vanjati, Girijan etc. are the synonyms for the tribes. Existence of tribal divorced from forest cannot possibly be conceived. The tribals have been living in the forest ecology and that has shaped their life and determined the kind of society they presently have. Environment, ecology with all its constraints determine the character of the society and prepare its dwellers to face all odd as well as enjoys all its blessings. Primitive communities among the tribes exist closely dependent upon the land or forest resources. The ecological system is a system of organism in a natural environment and in every community primitive or civilized what most importantly surrounds and influences the people are the traditions, sentiments, norms and aspirations that make up the mental life of the common people.

Thus ecology affects the social groupings and there is intimate connection between the ecological system and value and character of the people. For example, this fact is very well illustrated in the life style of the tribal and non-tribal communities living in Bihar and adjoining States like West Bengal, Orissa and Madhya Pradesh. History is full with instances of tribals heavily relying upon these hilly and rocky forested terrains as their impregnable fortress against British forces and their native retainers. Difficult rock and dense forest were their safest haven which enabled them and other tribal groups to carry on the depredations even in the presence of large police force in their very locality. The turbulence of wilds Korwas, Korkus, Khonds, Bhuinya, Nagesia, Kisans etc. were mainly because of their ecology which allowed them to show their stubbornness.

First Forest Policy (1861 - 1894)

In the last century the commercialization of timber did not take place in that measure which we witnessed in the first and second part of this century. The British administration did not like to take the risk of middling with the affairs of forest dwelling tribals in their hilly and forested areas. They had enough lessons from several revolts in Bihar and adjoining states. But British intention of taking over the management of forest in their own hand could not be delayed very much. In the year 1861 the Department of Forest was created and twelve years after, in 1873 the first legislation introducing the Government Policy and Programme came into force by the promulgation of the Indian Forest Act. In 1873 the Settlement of forest was undertaken and the entire forest areas in the country were classified into three categories:

(1) Reserve Forest (Class I & II).
(2) Protected Forest.
(3) Village Forest.

Under the new system, the traditional rights of the Tribal People over the forest was allowed to be maintained. The first Forest Policy was declared by the Government which allowed the forest dwellers to enjoy the rights to some extent but restricted from ownership. It stated:

(1) The sole object to which the management of forest is to be dedicated is to promote the general well-being of the country.
(2) The maintenance of adequate forests is dictated primarily for the preservation of the climatic and physical conditions of the country, and secondly, To FULFIL THE NEEDS OF THE PEOPLE, but these were subject to the following conditions:
   (a) Permanent cultivation should come before forestry.
   (b) The satisfaction of the needs of the local population at non-competitive rates, if not free overriding all considerations of revenue.
   (c) After the fulfilment of the above conditions the realization of maximum revenue should be the guiding factor.

Amendment Acts

This Act was followed by five different successive Acts passed, in the year 1901 Indian Forest (Amendment) Act, in 1914 the

Repealing and Amending Act, in 1920 The Devolution Amendment Act, in 1927 the Indian Forest (Amendment) Act. The last Act elaborately dealt with the different aspects of forest management and rights of the people in forest. It discusses at length Reserve Forests, .Village Forests, Protected Forests and control timber over forest and land not belonging to the Government and prescribed detailed penalties and procedures.

However, the forest under the jurisdiction of the Ruling Princes and Zamindars were excluded from the operation of these Acts. When Government of India Act, 1935 was passed, the forest became the provincial subject.

National Forest Policy 1952

After Independence, the declaration of new forest policy was made in 1952 which emphasized the preservation of national forest.

In Resolution No. 13/52/F, Dated the 12th May, 1952, the Government of India in the erstwhile ministry of Food and Agriculture enunciated a forest policy to be followed in the management of state forests in the country. However, over the years, forests in the country have suffered serious depletion. This is attributable to relentless pressures arising from ever increasing demand of fuel wood, fodder and timber, inadequacy of protection measures, diversion of forest lands to non-forest uses without ensuring compensatory afforestation and essential environmental safeguards and the tendency to look upon forests as revenue earning resource. The need to review the situation and to evolve, for the future a new strategy of forest conservation has become imperative. Conservation includes preservation maintenance, sustainable utilization, restoration, and enhancement of the natural environment. It has thus become necessary to review and revise the National Forest Policy.

The old and new forest policy when compared demonstrated that the "Privilege under the old (1894) became concession under the new policy (1952). The old policy envisaged the release of the forest land for the cultivation subject to certain safeguards. As for example in 1895 when demarcation of excess waste-land, jungle blocks of protected forest in Kolhan area of

Singhbhum district of Bihar was taken up, only 202 out of 911 villages under the settlement orders could be implemented. (Letter No 911 Foreign and 3916 L. R. dated the 18th February, and 14 August, 1895 respectively). In the rest of the villages it was found that the waste and forest were just sufficient for the reasonable requirement of the people or the excess waste and forest were broken up and scattered, its separation from the village was not practicable. In many of the blocks there were outland isolated plots of Rice and Gora cultivation. The Raiyots were induced to relinquish them, but they refused to give up the rice land. They were permitted to hold this land with distinct understanding that cultivation is not to be extended.

But the new policy of 1952 withdrew this privilege of primary cultivation of tribal cultivation over the forest land. The boundary pillars were erected after due notification of the area as the Reserve Forest, the affected people might or might not have objected. The old policy had left a margin for supply of the villager's requirements from the outlying areas of Reserve Forest. The new policy decided that there should be village forest for the purpose while the old policy did not touch the private forest of the tribals. The old policy did not prevent the grazing in the forest; the new policy brought it under the controlled grazing.

Para 32 of National Forest Policy, 1952 noted the following concessions to the tribals:

(1) Right to take water for agriculture purposes;
(2) Digging of wells or canals for agriculture purposes;
(3) Free grazing in open forests (under passes);
(4) Removal of stones and earth for domestic or agriculture use;
(5) Removal of timber, bamboos, seeds, canes etc. for construction and repair of houses and for agricultural implements;
(6) Collection and removal of head loads dead wood for domestic use;
(7) Collection of grass for cattle and for covering their huts;
(8) Fishing and hunting excluding the protected fauns; and
(9) Cultivation of forest lands.

Forest Policy and the Tribals

Before the British came into power, the forest dwellers and other population enjoyed freedom to use forest or exploit forest resources. There were, however, only the customary regulations of people's right over forests. Even during that period, certain traditional restrictions were there in the cutting of certain varieties of trees or in certain forests. Besides, as the population was small and vast forest areas were available, it did not really create any problem. When the British came in power, they realized the commercial value of the forests and tried to impose a number of restrictions on the use of forests and exploitation of forest resources. Before 1855, the different communities enjoyed freedom in this regard. But after 1855, restrictions were imposed on the rights of the different communities over forest resources. This was future modified and various earlier facilities were further curtailed and regulations were passed subsequently in 1878, 1894, 1927 and 1935 and in this process, the British gradually tried to bring the forest areas under their control. This would become clear if the different Forest Acts passed during the colonial period are analysed.

In August 1855, certain restrictions on the rights of the forest dwellers to conserve the forest were issued. They are "The sole objects with which State Forests are administered in the public benefit. In some cases the public is the whole body of the tax payers, in others the people on the track within which the forest is situated; but in almost all cases the constitution and preservation of a forest involve, in greater or lesser degrees, the regulation of rights and restriction of privileges of users in the forest area which may have previously been enjoyed by the inhabitants of its immediate neighbourhood."

Forest Act, 1865 – This covered mainly the management and preservation of forests. By virtue of this Act, the government was empowered to declare any land covered with trees as government forest. In order to establish government's control over forests, the administration was empowered to give punishment to anyone for the breach of the Act.

Forest Act 1878 - The forests were classified as:

(a) Reserved forests,
(b) Protected forests,
(c) Village forests.

Provisions were made to regulate or prohibit certain activities in the forest of different types and in this process, people's rights were further curtailed and more rigid control by the government was established.

Forest Policy, 1894 – The forest policy of the government was declared in 1894 and the major issues were:

(a) The sole object with which State forests are administered is public benefit. In general the constitution and conservation of the forest involve the regulation of rights and the restriction of the privileges of the user in the forest by the neighbouring population.
(b) Forests situated on hill-slopes should be maintained as protected forests to preserve the climate.
(c) Forests which are reservoirs of valuable timber should be managed on commercial lines as a source of revenue to the states.

Forest Act, 1927 – It was a more comprehensive one including those relating to the duty on timber.

Forest Act, 1935 – The subject of forest was included in the provincial list under this Act.

Infect, the British forest policies were mainly aimed at supplying of timber and other resources to British forest–based industries and commercial exploitation of forests was encouraged at the cost of the tribals in the name of national interest. Besides curtailing the various customary rights over forests, as the forest resources were mostly collected through contractors, the tribals became victims of commercial exploitation of forests. Observing serious exploitation in this process, Symington in late 1940, suggested that only the local people should be associated with the exploitation of forest resources. There is no doubt that the various measures taken during the British period restricted the rights of the forest – dwellers and the tribals, extended administrative control over forests and resulted into its increasing commercial exploitation.

Are we still following the colonial approach?

National Forest Policy 1952 - The New National Forest Policy was issued in 1952 and it was declared that the forest policy should be based on national interests and needs. It states that the use of forest "should in no event be permitted at the cost of National Interests. The accident of village being situated close to a forest does not prejudice the right of the country as a whole to receive the benefits of a National Asset. The scientific Conservation of a forest inevitably involves the regulation of rights and the restrictions of the privileges of the use, depending upon the value and importance of the forest, however irksome such restraint may be to the neighbouring areas." (FRI : 1981)

The approach of the government towards forest is also reflected in the National Commission on Agriculture (1976). It observed,

Free supply of forest produce to the rural population and their rights and privileges have brought destruction to the forests and so it is necessary to reverse the process. The rural people have not contributed much towards the maintenance or regeneration of the forests. (1976 : 25)

The Commission further recommended the functionally all forest lands should be classified into:

(1) Protected forests
(2) Production forests and
(3) Social forests

The Commission gave highest priority to production forests and the lowest priority to the social forests. Kulkarni has noted,

The Commission speaks of the needs of the rural population, but does not appear to take into account the Adivasi's need of minor forest produce .... Surprisingly, the Commission does not consider the Adivasi's need for timber for houses, leaves for thatching roofs, fruits, flowers and roots which are used as food, or seeds which are collected from forests to extract edible oil ……… The emphasis of the Commission is on man-made forests, also called production forests. (1988 : 14)

Infact, the Commission recommends that,

Each hectare of forest land should be in a position to yield a net income of many times more than what is being obtained at present. (1976 : 9)

The Draft Bill of the Indian Forest Act is mainly based on these recommendations of the National Commission on Agriculture.

Forest was a subject in the State List in the 7th Schedule of the Constitution of India till 1976. It was transferred from the State List to the Concurrent List by the 42nd Amendment in 1976 resulting diminution of powers of the states. Thus it is seen that National Forest Policy of the Government of India after independence is an extension of the old policy followed during the British period. This policy prescribes that the claims of communities near forest should not override the national interests and they should not be allowed to exploit the forest resources at the cost of wider national interest.

The different regulations and restrictions imposed on the forest - dwellers and tribals in the context of exploitation of forest resources on which their economy depends to an appreciable extent, virtually put them at the mercy of the forest department, particularly the lower level functionaries. The Forest Department, on the other hand, blamed the tribals for indiscriminate destruction of vegetation and wild life and suggested restriction on their rights over exploitation of forests. Report of the study team on Tribal Development Programme of the Planning Commission, while conceding the need to preserve the forests, pointed out, that the tribals have special claim over the forest and forest produces should be recognised. It was also reported that in many areas, even the collection of raw materials for their traditional handicrafts has become difficult. There is no doubt that the economy of the tribals and the forest - dwellers in particular is largely based on forest and forest produces. But in view of the various restrictions and regulations imposed from time to time, their rights to collect fuel, fodder and minor forest produces are very much restricted throughout India. Apart from the above problems, lack of social perspectives in the context of planning has also been emphasized by Roy Burman (1982).

The Forest Policy 1988 as well as the Forest Conservation (Amendment) Act, 1988 do not have a tint people statements that were conspicuous in the earlier documents. It does not accuse the people of destroying forests and does not state that they should be protected from them. However, the policy statement does not analyse any cause of deforestation. It only observes the critical situation and the urgent need to find a proper and suitable solution. It further observes that environmental stability should be the priority. A proper analysis perhaps would have resulted in some people-oriented solutions. It declares the needs of the industry, which is the main destroyer of forests, as national needs and the requirements of the tribals as people's needs. "Within the present power equation, national needs get precedence over those of the people. This has been the experience of the 1952 forest policy which gave equal importance to industrial and people's needs .... While this equality is appreciated, it should be clear from the analysis given above that till now revenue and industrial needs have got priority and the forest resources have, for all practical purposes, been transferred from the tribal and other rural poor communities to the corporate sector. Forests have ceased to be a community resource and have turned into a source of profit to industry and of revenue to the state. This has been done in the name of national development (Fernandes : 1992).

Like 1952 document, 1988 policy states as its objectives, the maintenance of environmental stability, conservation of the national heritage, increase in tree cover, meeting needs of the tribals, encouraging efficient use of forest produce and afforestation programmers. While this policy has more people - oriented statements, it still has the concept of exclusive state ownership of forests and mention industrial needs as national need.

Socially conscious environmentalist do not any more view the environment primarily as "beautiful trees and tigers." (Agarwal 1985) The control of these resources is transferred from the communities to the corporate sector to whom they are a source of profit. Naturally, there is a competition for the same resource between the powerful and powerless classes and one can well understand when the resources are depleted and competition for them is intense, who gets their benefit when access to them is unequal.

Persons conscious about the impact of the depletion of the natural resources on the traditional communities are equally concerned about the official response to this situation. The National Water Policy 1987 is a case where one also finds the trend of monopolization of natural resources by the powerful sectors. This is true for most other natural resources where the resources are transferred from the powerless classes to whom they are a mode of survival, to the powerful who treat and exploit them only as a source of profit. (Fernandes : 1992)

It is equally important to state that there has been a transition in the tribal culture from constructive to destructive dependence on forests. Tribals and other forest dwellers destroy forests to-day, through the sale of firewood (Agarwal : 1985). Thus to-day India has an estimated 2 to 3 million head loaders selling firewood. Also, grazing and shifting cultivation have today become sources of deforestation. The shifting cultivation cycle, for example, has today come down to 6 years in Orissa (Fernandes, Menon and Viegas : 1988) and to 10-12 years in the north-east. Policy makers, who look at only the present situation of destructive dependence, view it as opposed to the national needs of the industry and to environmental imperatives. The policy response is to protect forests from the people, particularly tribals. This response has taken various forms. In 1982 a draft forest bill attempted to give police powers to forest officials, to arrest offenders without a magisterial warrant. There have been other proposals to restrict or even abolish the "Concessions" given to the people as suggested by the National Commission on Agriculture.

The competition for scarce resources is, in reality, a conflict between powerful commercial – industrial interests and the powerless tribal communities. The forest policy only reflects this unequal situation. While all agree that there should be a people - oriented policy, one can assume that such a change will not be effected till the power equation is changed in favour of the tribals and other marginalised communities.

A New Forest Bill was proposed in 1995 to replace the Indian Forests Act 1927 by a new Act called "The Conservation of Forests and Natural Eco-system Act. "However, it was felt that the draft bill does not follow the Resolution on the National Forest Policy, 1988 in letter and spirit." The main salient features as laid down in the National Forest Policy Resolution, namely, "meeting the basic needs of the people - especially fuel wood, fodder and small timber for the rural and tribal people and maintaining the intrinsic relationship between forests and the tribal and other poor people living in and around forests by protecting their customary rights and concessions on the forests" have not been fully reflected in the official draft. The official draft did not take into consideration an important policy initiative of Joint Forest Management as reflected in the circular issued by the Ministry of Environment and Forests on 1st June, 1990. "Natural forests serve as a gene pool resource and help to maintain ecological balance. Such forests will not, therefore, be made available to industries for undertaking plantation and for any other activities. All must agree the last statement in the National Forest Policy Resolution that "forests are a national asset to be protected and enhanced for the well-being of the people and the nation." Again, the 73rd and 74th Amendments to the Constitution of India that suggested decentralization of powers has interesting implication in the context of forest management. It implies that only people can preserve and develop their natural resources and they must be encouraged to do so by assigning the management rights and duties.

REFERENCES

Agarwal, Anil and Sunita Narain, *The State of Indian Environment, 1984-85.* Centre for Science and Environment, New Delhi, 1985.

Chaudhuri Buddhadeb and A.K.Maity (ed.), *Forest and Forest Development in India*, Inter India Publication, New Delhi, 1989.

Fernandes Walter, Geeta Menon and Philip Viegas, *Forests, Environment and Tribal Economy,* Indian Social Institute, New Delhi, 1988.

Fernandes Walter (ed), *National Development and Tribal Deprivation*, Indian Social Institute, New Delhi, 1992.

Forest Research Institute, *100 years of Indian Forestry : 1861-1961. Vol.II,* Dehra Dun, 1981.

Gadgil Madhav and Ramchandra Guha, Ecology and Equity: *The Use and Abuse of Nature in Contemporary India*, Routledge, London, 1995.

Kulkarni S., Towards a Social Forest Policy, in Buddhadeb Chaudhuri and A. K. Maiti (ed), *Forest and Forest Development in India*, Inter India Publication, New Delhi, 1989.

National Commission on Agriculture, *Report of the National Commission Part IX, Forestry*, Ministry of Agriculture and Irrigation, Government of India, New Delhi, 1976.

Roy Burman, B. K., *Report of the Committee on Forest and Tribals*, Government of India, New Delhi, 1982.

2

STATE FOREST POLICY AND ADIVASI SELF-RULE IN JHARKHAND

SANJAY BOSU MULLICK

The indigenous peoples of Jharkhand demanded a separate province within the union of India for the fulfilment of their age-old aspiration for political autonomy and cultural identity. It was their millenarian dream that they kept compromising with the state at various phases of the struggle. 'Hundred years of anti-colonial struggle' that their fore parents fought for sovereignty had to continue for half a century more after independence only to receive a 'deformed child', that the Indian nation state gave birth to as her 28th province in 2000.The new millennium made a nasty joke of the millenarian dream of the Indigenous people of Jharkhand !

The folk poet sang 'we are being turned into Diasporas in our own land' years ago. Today the lament has become the cruelest reality for the indigenous people of Jharkhand. With a decreasing percentage of population, which is at the moment even less than 27 per cent, in the new state, euphemistically called Jharkhand, the IPs have very little hope of regaining political autonomy.

The ideology of the formation of smaller states (provinces) as opposed to the political recognition of cultural regions, under which Jharkhand was created, has served two purposes. Firstly, it divides the IPs of the Jharkhand cultural region and secondly, by having two smaller states out of a contiguous cultural region, Jharkhand and Chhattisgarh, with no dominant nationality, it makes it easier for the state to exploit their rich natural resources with comfortable ease. (Munda and Mullick, 2003, p. xvi). This is an intrigue, the plot of which is very similar to that of the colonial policy of 'divide and rule'. But it is not just a copy of the colonial policy; it is in fact the continuation of it to rule the 'internal colonies'. "The official Indian elite now constitute its own others in terms of continuing caste, class and gender differences within Indian society" (Jewitt, S., 2004, p. 58). Political scientists have noted that the state in post-colonial India has inherited a colonial attitude of mind from the British. (Nandi, Ashis, 1983 as quoted by Jewitt). For the British the Indians were the 'Other'; for the masters of the 'internal colonies', the IPs are the 'Others'.

The new state of Jharkhand has just completed five years of its existence on the 15th of November, 2004 dotted with series of violent conflicts between the IPs and the state administration. According to a conservative estimate police firing during this period has killed about 200 tribal persons. Several hundred people have been either jailed or rendered homeless and large-scale migration has caused a grave social frustration. This has been reflected in the just concluded Parliamentary and Assembly elections. The emerging crisis in the electoral politics has presented an unstable newly instituted government. It is a different story that despite being in minority the NDA could form the government by horse-trading. However, one must take note of the fact that only a little more than half of the voters officially caste their vote; and rigging was widely practice.

Thus the new millennium has unfolded another act of the drama of the IPs state relationship in Jharkhand and so to say in the whole country. However still this cannot be called as the last act of a tragic drama of the Adivasis of India as Nirod C. Chaudhury once observed on his famous book the Continent of Circe (Choudhuri, Nirod C., 1965). The struggle is continuing.

Post-colonial Forest Policy in Jharkhand

The basic cause of conflict between the state and IPs after independence is best exemplified by the post-colonial forest policy of the central and the state governments. The continuation of the Orientalist attitudes in the post-colonial power structure is remarkable. Instead of casting off the shackles of the British colonialism .... "many British ideas and methods of rule have been continued by post-colonial governments, especially as India has become more and more hooked into Western commercial and consumer systems. Instead of rejecting the modernizing policies of the British and replacing them with a more socially and environmentally sensitive forest management strategy, the post-colonial forest department actually embraced and enlarged on British policies"(Jewitt, pp. 71-72)

Before the beginning of the colonial era Jharkhand was really a 'land of forest'. The land under swiden cultivation (daha/ dhya) was very common. Only in some pockets plough agriculture was introduced by the medieval 'jungle' states. But in both the cases the motion of land belonging to the king as the 'lord' paramount of the soil was absent. Community enjoyed total control over the land and the produce of the land. Forest was an integral part of the economic system. Forest were dense and full of wild animals. According to the belief of the indigenous people the real owner of the bir-buru (forest and mountain) and ote –hesa (land and cultivable fields) is the Supreme Being –the creator of the universe. However in those places where state formation started out of non-indigenous matrix the people were forced to pay a sort of tribute to the kings, not against the land that they possessed but for the protection against any external aggressions. In the case of the largest of such states, the state of Chhotanagpur, the tribute was called chanda (donation). People never paid it regularly.

The Diwani of Bengal, Bihar and Orissa was given to the British East India Company in perpetual grant by Sah Alam, the Mughal Emperor of Delhi on the 14th of August, 1765. The grant of Diwani was a grant of right to collect revenue and to exercise jurisdiction, powers in all civil and financial causes. However the Mughal agrarian system was replaced by the strike of a pen in 1793 when all the independent and free cultivators were turned into tenant and the village land was turned into the private property of the newly formed class of landlords. Jharkhand which was hitherto considered to an independent land only loosely connected to the larger political power was made a part of Bengal. Jharkhand was officially brought under the purview of the permanent settlement initially only on papers. In reality it took 100 years for the British to establish its rule in Jharkhand. The British confrontation with the Indigenous peoples began in the eastern fringe of the 'Jungle Mahal' as the Mughals called it or the Bengal woodland as it was christened by the British by the beginning of the 19th century and it ended with the suppression of the Birsa Rebellion in 1900. Chhotanagpur came under British rule in 1818. The objective of the political control was revenue from land and timber from forests. But that was not possible without pacification of the rebellious peoples. Therefore, the British policy was marked by three elements, suppression of protests, sedentarization of shifting cultivators and 'conservation' of forests. A large corpus of literature is available on the history of the protest movement but very little attention has been paid to the study of other two element of British policy in Jharkhand.

"Colonialism made a distinction between public and private domain. While environment was placed in the former, agriculture was made a private sector activity. The partitioning of landscapes and social spheres came to characterise the large aspects of modern state formation."(Shivaramakrishnan, K., 1999, p. 80) In this process tribal places were made, not only by curbing practices like shifting cultivation but by assigning them a specific terrain which was designated the place of tribes unredeemable from 'backward agriculture.' (*ibid*, p. 81). By the 1870s, the south-western part of woodland Bengal the emerging tribal heartland – was taking shape as a landscape where two element of policy converged. One element worked to hasten the gradual transformation of wild lands and wastelands into an ordered terrain of fields and groves. The other element was the policy of forest conservation." (*ibid* p. 86)

The initiative taken under the Indian Forest Act, 1865 to replace the community managed forests by state management reached its goal when the process of state usurpation was consolidated in the Indian Forest Act, 1878. The reservation of large areas of forest by the colonial state under Reserve, Protected and Private category thus empowered the states immensely and consequently diminished the rights of the ecosystem people to a large extent. This changed the traditional pattern of resource use and made timber an important commodity, which in turn fundamentally altered forest ecology (Gadgil and

Guha 2000, p. 85). The dominant forest structures were commonly changed from all aged diverse and naturally regenerating jungles to simplified even aged monocultures often dominated by sal, teak and pine. (Jewitt S. 2004, p. 62)

By the end of the 19th century the British rule in Jharkhand accomplished two important tasks, the suppression of protests and sedentarization of the IPs. This achievement was followed by the adaptation of paternalistic and isolationist policy, which was reflected in the tenancy acts and administrative policies. Chhotanagpur Tenancy Act of 1908 and Santal Pargana Tenure Act of 1912 were framed and IP dominated districts of the region were put under 'partially excluded areas' with a simple administrative system. This was necessary to keep the region under direct control of the Raj and to keep the IPs away from the growing anti-colonial movement as the discovery of coal followed by huge deposits of iron, manganese, bauxite etc. made the region the most lucrative land of the colony. This was true to the other IP habitats of the country as well.

After independence the Indian ruling elite under caste-class combine, categorised as 'omnivores' by Gadgil having satisfied its hunger of power engaged itself in satiating its thirst of wealth. Semi-feudal structure of Indian society refused to accept land reform (1956) and sharing of power at powers at the village level under Panchayati Raj (1957). 'National interest' was the 'Log in name' and 'development' became the password to have a smooth access to the natural resources of the country for their inhuman exploitation and criminal expropriation. Public servant, police and politicians formed a nexus to this end. Where colonialism left of development to cover (Kothari, R. 1988, p.143 as quoted by Jewitt). World of artifacts devastated the natural world. While Dams and Mines displaced millions of peasants and tribals, destruction of forests caused hunger and destitution. The swelling multitude of 'ecological refugees' (Gadgil, M., 2000), who constitute about one third of Indian population, were turned into sweet labours and treated as the dirt of development. Development in Jharkhand took place the threat of guns. Behind the repression that accompanies development lies a perception of the Adivasi peasantry as physical obstacle in the drive to gain full access to land, raw material and natural resources (Devalle, S. 1992, pp. 104-105).

Jharkhand has been the largest producer of ecological refugees. More than 300,000 people have been displaced under the development projects; 90 per cent of them are the IPs. Ecological devastation has completely broken the chain of food security. According to the official estimate 10 per cent people of Jharkhand today suffer from hunger frequently and 2 per cent suffer from chronic hunger.

Development projects alone are usually blamed for the ecological destruction and consequent devastation of the tribal society. The role of the forest department in this gold rush at the cost of the human and nature was neither studied nor exposed till recently when the confrontation between the forest dependent people and the forest department could not be suppressed any longer in the central Indian states and the foot hills of the Himalayas. Chipko, Jharkhand and Narmada movements exposed the highhandedness of the forest departments and the rampant corruption in them besides challenging the forest policies of the government. Forests continue to be the largest zamindari (fiefdom) of the state and the forest officials act like despotic landlords.

In Jharkhand the history of 'Crime against Ecology' may be studied in three phases. The first phase began with the British occupation of forests as eminent domain and clearing of flora and cleansing of fauna for the production of agricultural surplus. In the second phase British conservationist policy was challenged by the Zamindars during the 2nd world war and by the forest department after independence till the central government tried to revive the policy of 'conservation' in the 80s by the enactment of 1980 Act. And the third phase began after that with a series of laws, resolutions and directives aimed at completing the unfinished job of the colonial masters of legally severing all relationships between forests and forest dependent peoples, especially the 'adivasis' or Indigenous Peoples.

The contradictory nature of much British policy-making was exemplified in the conflict between normalizing/dominatory aspects of British rule, and a paternalistic desire to protect the savage but noble adivasis (Jewitt, 2004, p. 64). The Zamindars

(Landlords under permanent settlement) always hated this and the nationalists termed it as a Christian ploy to convert simple hearted adivasis. The nationalists stood for assimilation as opposed to British unstable policy of isolation. The contradiction and complexity were inherent in both forest and tribal policy of the Raj. British tried to achieve an achievable goal: the forests were sought to preserved by alienating the Adivasis from it and the adivasi identity was promised to be protected by severing their symbiotic relationship with the forest.

The Zamindars and the forest officials tool full advantage of the unstable political conditions before independence, marked by quit India movement and Second World War, and unsteady tribal and forest policy frameworks after it for about half a century beginning from 30s to the beginning of the 80s to convert natural woods into artificial wealth. The British government during the fag end of its rule decided to take over the village forests from the custodianship of the Zamindars by enacting the Bihar Private Forest Act of 1946 (III of 1946). It was 're-enacted as Bihar Private Forest Act, 1947 (Act IX of 1948) with certain modifications' after independence. When Zamindari was abolished and operation of the Land Reform Act (Bihar Act XX of 1950) was initiated all these forests became vested in Government and were constituted as Protected Forests (PF) under the Indian Forest Act of 1927. However, that turned out to be a boon in disguise for the official of the forest department. They cleared what was left out by the Zamindars as a gift of independence and reward to their loyalty to the British Raj.

In its quest for quick currency the Forest Department in collusion with the revenue department declared the Mundari Khuntakatti villages as Zamindari Villages and notified the forests as Private Protected by overtly violating the provisions of the Chhotanagpur Tenency Act, 1908 under which the Adivasis belonging to the Munda community enjoy a special status. However, the Mundari Khuntkattidars lodged a strong protest against this injustice on the ground that the Mundari Khuntakatti tenure was quite different from the Zamindari tenure. They were neither tenant nor tenure holders and they collectively owned the forests within the boundary of the village. However, it yielded no positive results and the Forest Department virtually took away the Mundari Khuntkatti forests from the Mundas with a plea of managing their forests 'scientifically' and a promise of sharing the profit that they never kept. In 1984 the Roy Burman Commission visited these villages and found to its utter disbelief that miles after miles forests were denuded mercilessly by the forest department.

During this phase in the 70s Jharkhand witnessed a fierce conflict between the Adivasis and the Forest Department. Areeparampil noted four basic reasons for this. They were, encroachment by the Forest Department on adivasi villages and on their customary rights on forests, exploitation and harassment of adivasis by forest official, commercialization of forests and resultant pauperization of the adivasis and looting and destruction of forests by contractors in collusion with forest officials (Areeparampil, 1992). Conflicts of the same nature in other parts of the tribal India revealed their long lineage. Local opposition to commercial forestry dates from the earliest days of state intervention (Gadgil and Guha, 2000, p. 85). Scholars observed the political edge of such movements. In Jharkhand it was combined at the outset with a radicalization of the movement for a separate Jharkhand state (Sengupta, 1982, Devalle, 1992, Corbridge, 2004) and elsewhere in the central tribal belt the issue was actively taken up by the Revolutionary Maoist groups (Gadgil and Guha, 2000, p. 85). In Jharkhand adivasi movements for land and forest were always a part of their autonomy and identity movement. They not only criticized the anti-people policies of the government but at the same time presented the demand of autonomy as the means of resolution of the crisis.

The government, however, wanted to deal with the situation by passing a more stringent law that would give the forest department absolute control over the forest land. The Draft Forest Bill of 1980 invited a country wide opposition and finally the state had to change its strategy. Orientalism came back in a new form: the Joint Forest Management. This third phase of our study reveals the revival of the colonial dichotomy. But as the colonial policy of tribal paternalism and forest conservation had an ulterior motive the JFM was also conceived with a selfish intention. The Forest Conservation Act, 1988 is often seen as a departure from the 1980 Act, a revolutionary step forward. In fact it complements the Act. The call for people's participation in forest conservations is not a gesture of goodwill to the people but a strategy to use people whom the other act deprives of their forest rights under compulsion.

JFM policy was adopted by the Bihar Government way back in November 1990 when a large tract of forest land was already degraded and the forest department became completely alienated from the people. (Government of Bihar admitted way back in the 70s that out of 29,232 sq km of declared forest 10,000 sq km was degraded (A Draft Perspective Plan, Bihar – 1978-89). Presently, the Jharkhand Government admits that there is only 18-19 per cent of the total state land under forest cover (Vision 2000, Jharkhand Government) as opposed to the conflicting claims of 27 per cent by the Forest Department).

Therefore, it was not surprising that the progress of the programme staggered with a ridiculously negligible average of 17 JFM community organizations formed every year. However, the expectation was quite high. In an important policy shift, the Indian National Forest Policy of 1988 underscored the need to involve local communities in the management of forests: and regeneration of forests and alleviation of rural poverty were the expected outcome (Sanjay Kumar, 2004). With a view to assuring the involvement of the forest peoples 33 per cent of the total net profit was promised to be shared with them. The programme was hailed as a revolutionary new scheme (Jewitt, 2004).

After the formation of the Jharkhand state, the new state government reinitiated the project. The Government of Jharkhand has a new Joint Forest Management Resolution dated September 27, 2001, that supersedes the Government of Bihar Resolution on the same issue (GoB Resolution No. 54/90- 5244 V.P. dated 09.11.90). The main objective of the resolution is to establish joint forest management committees for the protection of Reserved and Protected Forest of the state and for the betterment of the villagers living in these forests.

The Jharkhand Resolution begins by accusing the forest dwellers as the destroyers of the forests. Then it allures them by promising 90 per cent of the net profit made out of the sale procedure of the forest produces including timber and demands responsibility of protection and regeneration of forests in return. Forest Protection Committees will be formed in the villages, the portfolios of the secretary and the treasurer will be held by the Forest Department. The members of the committee will be given the policing power and authority to collect fines for small offences. A detailed analysis of the Resolution makes it clear that the sole purpose of it is to re-establish a centralized power in the forest with the help of the forests dwellers. 90 per cent profit sharing is a hoax because it is the net profit, which will be taken into account and after the colossal expenses of the Forest Department, it is anybody's guess how much would be left for distribution to the people. Moreover, people will not be deciding the price of the forest produces and will have no say in the money management. It violates the Chhotanagpur Tenancy Act, 1908 and the Panchyati Raj (Extension in Scheduled Areas) Act, 1996. The most dangerous consequence will be a division within the village and between neighbouring villages on the issue of stopping people's access to the forest.

Indigenous Peoples' struggle for autonomy and identity throughout the colonial and postcolonial period demanded the state recognition of the integrity of their culture and nature. As forest is inseparable from their existence, the tribal and forest policies of the state could not be framed on incompatible premises. However, this demand was never honoured by the state. The present Forest Policy and the Draft National Policy for the Tribals (DNTP), prepared by the Government of India in 2004, are no exception to this rule. While the former holds a conservative ideology and promotes eviction and isolation of the IPs from the forests, the later does not even talk about the forest ownership rights of the people.

The Panchayati Raj Extension Act of 1996 in Scheduled Areas denies the ownership rights of the IPs in the village forests. Moving a step further, the Jharkhand state Panchayati Raj Act of 2001 has taken out even the rights over the Minor Forest Produces from the jurisdiction of the Gram Sabhas, which was granted in the Central Act.

World Bank Forestry Projects in Jharkhand

The World Bank has also endorsed this policy of separation of the IPs from the forests by accepting the JFM as the guiding principle of its involvement in forestry programme in India. The Bank planned to extend its involvement in forestry programme in India. The Bank planned to extend its involvement in the forestry sector of Jharkhand after covering other tribal dominated Indian States in 2002. It called a consultation meeting in Delhi to this end with civil society representatives. The Bank pointed out, 'Recognising that some of the poorest, most marginalized and under served communities in India live on the forest fringe', the Bank is committed to finding a way to support their development (Poverty Reduction and Forestry in India. Ambassador Hotel, New Delhi, September, 2002).

In November 2003, World Bank official observed the annual meeting of Jharkhand Save the Forest Movement where interventions by tribal participants clearly articulated their opposition to the state JFM policy and the World Bank project-positions affirmed in the Khunti Declaration which stemmed from the meeting. The fact that the Bank has since pushed ahead in 2004 and approved preparation of the project and support for a pilot JFM project suggests that the position of many tribal leader has been disregarded or at the very least is being treated as unimportant.

Tribal peoples of Jharkhand have rejected the State Government's 2001 JFM policy in a series of memoranda and statements over the last two years. Despite this popular opposition, the Bank's own Project Information Document (PID) explains that the primary purpose of its so-called Participatory Forest Management Project will be to support the implementation of the Jharkhand government's JFM policy which it describes as "progressive". Tribal representative point out that the Bank is planning to back a forest policy that is not acceptable to tribal communities who live in and depend on Jharkhand's forests. Under a popular pressure the Bank has recently invented a new term called the Participatory Forest Management to replace the Joint Forest Management; the content, however, will remain the same.

The struggle continues. Forest rights movements in India, in general and Jharkhand, in particular, have become a coordinated effort to force the newly elected central government to recognise the rights of all the forest dependent peoples in forests. The United Progressive Alliance mentioned its promise to redress the grievance of the forest dwellers in its common minimum programme before the elections and now it has honoured its commitment and is prepared to present Scheduled Tribe and Forest Dwellers (Recognition of Forest Rights) Bill, 2005 in the Budget Session of the Parliament. But as we know, there is always a gap between the lips and the cup. Good intensions in the state processes very often are frustrated by the state mechanism. As it happened in the case of the Tribal Panchsheel of Nehru and, in recent times, the Bhuria Committee Report for the Panchayati Raj (Extension in the Scheduled Areas) Act, 1996 there have been strong apprehensions among the people that the colonial remnants in the power structure would fight to the end to stall or deform this Bill as well.

REFERENCES

Areeparampil, Mathew, Forest Andolan in Singhbhum in S. Narayan(ed.), *Jharkhand Movement: Origin and Evolution*, 1992.

Choudhuri, Nirod C. *The Continent of Circe Chatto & Windus, London*, 1965.

Corbridge, S., Stuart Corbridge, Serah Jewitt and Sanjay Kumar, *Jharkhand.* Oxford, New Delhi, 2004.

Devalle, Susan. B.C., *Discourses of Ethnicity*: Culture and Protest in Jharkhand, Sage, New Delhi, 1992.

Gadgil, M and Guha, R., *Use and Abuse of Nature*, Oxford, New Delhi, 2000.

Guha, R. *Unquiet Woods*, Oxford, Delhi, 1989.

Jewitt, Sarah, Europe's Others? Forestry Policy and Practices in Colonial and Post-Colonial India in Corbridge, Stuart *et al.*, Jharkhand, Oxford, New Delhi, 2004.

Munda, R.D. & Bosu Mullick, S. (ed), *The Jharkhand Movement*, 2003.

Sanjay Kumar, Stuart *et. al.*, Jharkhand, Oxford, New Delhi, 2004.

Sarin, Madhu, *Gender Relation in Forest Societies in Asia*, (ed.) Kelkar, G., Nathan, D. Walter P., Sage Publication, New Delhi, 2003.

Sengupta, N. (ed), *Fourth World Dynamics of Jharkhand*, Author's Guild, 1982.

Shivaramakrishnan, K., *Modern Forests*, Oxford, New Delhi, 1999.

3

IMPLEMENTATION OF FOREST LAW
A Case Study in East Singhbhum of Jharkhand

ASHOK KUMAR SINHA

Introduction

After the bifurcation of Bihar and formation of Jharkhand, separate forest laws have not been formed for the newly born state. Reviews of literatures indicate that the existing forest laws of undivided Bihar have been implemented here in toto. Because of this, most of the loopholes of the forest laws have been inherited by Jharkhand. Hence, their effective implementation has been a difficult task. The east Singhbhum district of Jharkhand state is witnessed by forest coverage of 1049.40 sq km area. This forms approximately 30 per cent of its total land area of 3533.35 sq km. The world famous Dalma range is covered with green carpet consisting of broad spectrum of plant species, providing an enormous genetic base for their sustainable utilization. The Jaduguda hill range also forms an extensive patch of forest which is rich in plant diversity. Efforts have been made in present communication to trace the history of legislation aimed at preventing the decreasing forest area of East Singhbhum of Jharkhand.

Framing of Forest Laws : A Historical Perspective

The Indian Forest Act – 1927

This was the first attempt made by British India to insure the attainment of the constitutional goal of protection and improvement of the natural wealth and the environment and safeguarding the forest. It was also aimed at the preservation of fauna and flora, some species of which are getting extinct or becoming endangered.

The National Forest Policy: (1952, 1988)

Although the National Forest Policy was framed by the British Government in the year 1894, it was made more relevant after independence. It was formulated on the basis of certain needs of the country, they include prevention of denudation of forests, erosion of forest land and establishment of tree-lands. It is also aimed at sustainable supply of timber and other forest products required for defence. Efforts have been made to classify the forests through this policy. Four types of functional classifications were made viz. (a) Protected forest (b) National forests (c) Village forests and (d) Tree-lands. The national forest policy was reviewed in the year 1988 due to the fact that over the years forests in the country have suffered serious depletion.

For the management of state forests schemes were projected to meet the growing needs for the essential goods services which the forest provides. Emphasis was laid on the necessities to enhance the forest cover and productivity of the forest through the application of scientific and technical inputs.

Forest (Conservation) Act – 1980

This act was brought to ensure judicial intervention in case the administration fails to take preventive, remedial and curative measures to preserve the ecological balance.

Forest Produce Regulation of Trade Act – 1984

This act is also known as Bihar Forest Produce (Regulation of Trade) Act, 1984. This deals with the restriction on purchase or transport of specified forest produce.

The Environmental (Protection) Act – 1986

This is an act to provide for the protection and improvement of environment and for matters connected therewith. This law extends to whole of India. Hence, our state Jharkhand also comes within the purview of this act. Different technical terms have been defined for their specific uses viz. Environment, Pollutants, hazardous substances, occupier etc. According to this act, the central government shall have the power to take all such measures as it deems necessary or expedient for the purpose of protecting and improving the quality of the environment and preventing environmental pollution. As per this act, the government has the power to establish one or more environmental laboratories to monitor the level of the pollutants and other degrading factors.

Bihar Saw – Mill (Regulation) Act – 1990

This law extends whole of Bihar (and, therefore whole of Jharkhand). It came into effect from May 1990. According to this act, no person shall establish a saw-mill or a saw pit except under the authority and subject to the conditions of a licensee.

Types of Forests and their Significance

Forest may be (a) Deciduous, (b) Semi deciduous, or (c) Ever-green. However, majority of forests of Jharkhand are semi-deciduous. Most of the trees are highly valuable with respect to their timber and medicinal importance. Plant diversity of East Singhbhum is witnessed by the presence of tree species viz. Babool, Bel, Siris, Kathal, Neem, Kachnar, Chironji, Palash, Paasu, Sisam, Kendu, Amla, Gamhar, Kurchi, Mahua, Aam, Karanj, Saal, Bija Saal, Muchkand, Semal, Jamun, Imli, Arjun, Bhera, Haritaki, Ason, Ber, etc. Different shrubs include Chirchiri, Chiraita, Shialkanta, Bans, Vajradanti, Akwan, Datura, Van tulsi, Amari, Basak, Putus, Baghnochi, Gajar ghas, Jaundice-patta, Arandi, Kans, Makoi, Sadabahar, Baigna, Chota gokharu. In addition to these plants various grass species and aquatic plants are also found in clusters.

The forest vegetation of East Singhbhum has not yet been documented systematically. Some works have been done by forest officers (Deo, 2000) posted in the Dalma range. Many tree species have not been investigated with reference to their genetic potential and other values. Wide-spread anthropogenic activities are responsible for elimination of a significant number of plants before we come to now about their importance.

Causes of Forest Degradation

Among the various causes, the illegal felling of trees and their unauthorized trading is the major one. The illicit cutting of trees of different species has denuded the forest. The local people use felled trees for their use and the surplus materials are sold in the local market of Patamda, Chandil, Pardih, Mango, Jamshedpur and Sonari etc. Due to lack of communication system and limited staff, weak patrolling system, the administration finds it very difficult to check the illegal cutting of trees. Poverty of local people, their connivance with dealing forest personnel, is also responsible for forest degradation.

The wide-spread practice of forest hunting, called Sendra, and other illegal poaching happens to be one of the major reasons of forest degradation. The tribal hunt is carried out during April and May of every year. The Dalma range and Jaduguda Area have been experiencing enormous biological pressure due to tribal hunting. Domestic livestock, grazing and cutting of tender branches for use as fuelwood also lead to forest degradation. According to a recent report (Deo, 2000), most of the forests of Dalma range are vulnerable to the forest fire. Most commonly, it is the ground fire that swipes away the fallen dry leaves, humous and other organic matters which together form the forest land productivity. In this way fire reduces the productivity of soil and rate of infiltration for nutrients which directly affects the forest growth.

According to this report, Dalma hill is witnessed by illicit distillation in local distilleries called "Bhatti", which go on unabated all round foothills of Dalma. These bhatties are operated by the use of fuel wood collected or harvested from the adjoining area of forest. They also prepare "Haria" (rice beer) by the use of fuel wood harvested from forest.

Legal Status of Forest Protection in East Singhbhum

The Wild Life (Protection) Act–1972 is still in force with reference to Dalma sanctuary. According to this, provisions are made to carry firewood and small timber by local people for their bona fide purposes only in the protected forest. After declaration of this sanctuary, the working of forest in the core was banned from the very beginning of its creation. In buffer zone, it was banned afterwards and as per today. The operation is being done after the transfer of whole area under the administration during the period 1991-92. All villages inside and outside the sanctuary are made revenue villages. As far as grazing rights are concerned, there are prescribed and set rules for grazing before the area was declared as a wild life sanctuary. In the present scenario rules and regulation needs are to be prescribed to control grazing.

Implementation of Forest Laws and Protection of Intellectual Property Right

After the implementation of GATT (General Agreement on Trade and Tariff) it has become extremely essential to protect each and every biological resource as intellectual property. Since the vast resources of plant kingdom provide enormous wealth for exploiting as medicines etc. efforts should be made to bring these resources under the intellectual property right and patent laws. The patenting of Basmati rice as "Taxmati" by USA has become an eye-opener for the law-makers of the country. Hence, such types of laws should be enforced to protect and conserve our forest resources. Certain Gymnospermous trees have been found to produce chemicals with anticancer properties. Such trees should not be allowed to be harvested unnecessarily.

In the most industrialized countries, plant varieties are subjected to intellectual property right. It is designed to promote the protection of diversity of whole ecosystem or unmodified plants. However, the evaluation, use and maintenance of plants may be encouraged directly by the patent system when constituents of the plant have a commercial value.

At present, we do not understand fully the implication of the extension of intellectual property right to plant genetic materials and crop variety. This may have far reaching impact on maintenance and preservation of plants genetic resource (Gupta, 1995).

Conclusion

The newly created state of Jharkhand has enormous botanical wealth. It is also regarded as one of the mega diversity areas of the world. The documentation of forest flora is the need of the hour. The East Singhbhum of Jharkhand should get proper governmental attention to conserve the fast depleting forest area. Existing forest laws should be strictly implemented and new laws should be framed where old ones are no more effective. People's participation is most desirable in the implementation of laws.

REFERENCES

Bandhu, Desh, Garg, R.K., *Social Forestry and Tribal Development*, Indian Environmental Society, New Delhi, 1986.

Deo, Niranjan Prasad, BFS, Management Plan of Dalma Wild Life Sanctuary; Department of Forest and Environment, Government of Jharkhand, 2000.

Ghate, S. Ruchas, *Forest Policy and Tribal Development*; Concept Publishing Company, New Delhi, 1992.

Gupta, P.K., *Elements of Biotechnology*, Rashtogi and Company, Meerut, 1995.

Sharma, Atul, *Paryavaran Aur Van Sanrachna Samashya Evam Samadhan*, Taxila Publisher, New Delhi, 1991.

Sikligar, Dr. Punamchand, *Van Evam Adivashi Samajik Jeevan*, Shiva Publisher Distributor, Udaipur, 1994.

Trivedi, Priya Ranjan, (ed), *Environmental Protection Law*; Published by Indian Institute of Ecology and Environment, New Delhi, 1995.

Virotam, B., *Jharkhand Itihas Evam Sanskriti*, Bihar Hindi Granth Acadami, Patna, 2001.

4

THE COLLECTIVE WISDOM AND EXCELLENCES RELATED TO WORLD VIEWS ON FOREST, BIODIVERSITY AND NATURE-MAN-SPIRIT COMPLEX OF THE INDIGENOUS PEOPLE OF EASTERN INDIA

PASHUPATI PRASAD MAHATO and KSHIROD CHANDRA MAHATO

The indigenous and tribal people of eastern India are mainly found in the different states of India. The Mundas, Ho, Oraon, Santhal, Kudmi, Kharia, Lodha Savar, Pahariya etc. ethnic groups who claimed themselves each a separate ethnos in their own territory are distributed in the states of West Bengal, Orissa, Jharkhand, Bihar, Assam and Tripura.

The indigenous or autochthons of this geo-cultural politico regions were termed by the Aryan invaders, who subjugated their territories of settlement through the ages. They were termed by the nomenclatures, which are other than *Men*. In North-East India, the indigenous people were termed as *Naga* or the naked people. But the meaning of *'Naga'* in their own language are the *people of heaven*. The most ancient social civilizations of our country were the Bodo and Naga civilizations of North-East, who called themselves as *Borok* meaning man and *Naga* meaning also men. On the other hand Munda, Ho, Santhal, Birhor etc. called themselves as *Horo-Ho* and *Ḥor* respectively. The Kudmi-Mahatos who were the most ancient people were also known as Kutumbin and claimed themselves as Kudmi-Hor. But as the cultural aggression, language, violence, and territorial subjugation became rampant, the social civilization of the Hor-Mitan or Horo-Mitan in present day Jharkhand, Orissa and

West Bengal the ancient 'Kherowals' became fragmented into different ethnic groups and now claiming themselves each one as separate ethnos or ethnic groups.

Bosu Mullik, Jayadas, Akkara and Jayadas (1993) remarked – 'A sanskritised section of the great Shavaras in Orissa is called as *Suar (Swain)*. The Kurmis in Bengal are called as *Chuars* meaning barbarians. The name *panias* in South India stands for 'Slaves'...... In formative periods of the emergence of the Aryan civilization the Aryan rulers called the subjugated people as the *dasas* (slaves). The *Kutumbins* were the *dasas* as recent researches on Sanskrit texts were explored by Ashok Kumar Mahato (2002). "The rest of the indigenous people who resisted the aggression of the Aryans were known as Jana and their settlements were called 'Janapada'. The indigenous people who fought against the invaders tooth and nail were called as *Rakshasa*, meaning cannibals and *Dasyu* meaning bandits". "The history of the indigenous people of India has been the history of stubborn resistance to the colonial onslaught of the Aryans and the Brahminical Hindu States, Muslim invaders and their empires, East India Company and the British imperialism and finally of the advanced nationalities and the 'Internal colonialism' of the modern Indian states".

The first mythological and symbolic expansion of the Aryans was the subjugation of the–*Khandavban*, where *Takchchak* was the king of the Nagas (The totem symbol), where the ethnic groups like *Paskki, Nisad, Kirat, Hasti* and *Kukur* etc. used to live. When Dhritarashtrya granted 'Khandavban' to the Pandavas for their settlement, the horror and butcher of the indigenous people started. But we the elites of the indigenous people never reinterpreted the Khandav myth.

It was a myth to legitimize and to invade the collective psyche of the indigenous people through the so-called Indian epic Mahabharata. Thanks to Dr. Dipak Chandra who explored and reinterpreted Agnigarva Khandav. He wrote – To the inhabitants of Khandav, the Pandavas were the outsiders. The authority and control over 'Khandavprastha', by them was never accepted by the indigenous people of the area.

Since then outsider's concept came to the indigenous people, which is at present termed as *Dikku.* But by the able leadership of Takchchak the entire Nag people combined and resisted the Pandavas.

Now who are Nag People? Nag is the most respected totem of the indigenous people of India. Now we must understand the anthropological meaning of the totemic and titular, through both the words are related to common ancestors. K. S. Singh (1992) pointed out– "Anthropological definitions of clan and lineage are clearcut, one relates to a mythological ancestors and the other to a genealogically traceable ancestors. The term 'vansha' has been used rather loosely for both categories. Lineage is replacing clan in popular usages. By titular we mean the assumed ancestors who were sages like Bhardawj, Basistha, Kasyap etc mainly adopted by the upper castes of India".

But a totem may be a fruit, an animal, a bird, a fish, an insect, a tree, etc, which they regard as the most sacred, which should not be killed or eaten. Those objects or animals or birds should be protected at any cost. These totemic based world views of the indigenous and tribal people which flourished the social civilization of the *Horo-Mitan* or *Ho-Mitan* or *Har-Mitan* in *Radh-bhum* or Jharkhand is unique to protect the bio-diversity of forest based societies of Jharkhand, Chhattishgarh, West Bengal, Orissa and Tripura. Take the cases of the Jharkhand area –

The Mundas have got the totem Soi meaning sol fish, Nag (serpent), Hassa (goose), Kerketa (a kind of bird), similarly the Santhals have the totem Murmu (a forest based wild cow), Baske (a serpent), Chande (a lizard), Boyar (a fish), To an Oraon, Kachap (tortoise), Kujur (plant), Tirkey (a bird), Kispotta (a wild bear), Bakhla (a plant), Bagha (tiger), similarly for a Ho Hansda (a wild goose), Bage (tiger), Jamuda (Spring), Sandil (a bird), Tiyu (fox) and with the Kudmi of Jharkhand Karuar/Kaduwar, (wild buffalo), Kairoyar (banana), Hindowar (a small fish), Chiluar (kite), Nag (serpent), Lakrawar (wolf), Kheruwar (Khorgose), Dumriar (pig), Bansiar (bamboo), Musayar (rat), Kachchuar (tortoise), Baghowar (tiger), Bakuar (crane), Chiluar (kite), Pakruar (a fruit), Mutruar (spider) etc. are very interesting. Each of the ethnos who termed themselves as Munda, Ho, Santhal, Kudmi etc. never eat, or kill animals, birds, fruits etc. On the other hand, during the common lineage festival known as *Dharm puja,* every lineage member has to undergo rituals of fast to worship his own totem.

Those observances of sacredness of their totems are unique only with the indigenous people of Jharkhand. Thereby they not only protect the equilibrium of the biodiversity, but also continue the wisdom and collective excellence related to forest and their protection of biodiversity. Thus when ILO-convention 1957 (no.107) at Geneva 1-10 September, 1986, meeting of experts on Revision of the indigenous and Tribal population gave the definition- *"Indigenous communities, peoples and nations are those which having a historical continuity with pre-invasion and pre-colonial societies that developed on their territories, consider themselves distinct from other sectors of the societies now prevailing in those territories, or parts of them. They form at present non-dominant sectors of society and are determined to preserve, develop and transmit to future generations. Their ancestral territories and their ethnic identity as the basis of their continued existence as people, accords with their cultural patterns, social institutions and legal system."*

II

The forest, since time immemorial, has been depicted in the indigenous people's folklore and there by depict collective creative excellence.

Eminent historian of Jharkhand Late Sunil Kumar Sen collected the following Ho Song, which was jointly translated with Prof. Dr. Sudhir Karan into English.

Singh – Subah – Resinghbonga Seba Kate Disum Keda
Singh Disum Nutun Te Ko Nutun – E- Tade
Bir – Bure Tandu Kate Kula-Buru Har Kate
Bir – Ko Tala –Mala Re Ko Disum – E- Keda
Dir Daru Aming Kate, Gora Loyong Tandi Kore
Bera Nari Ko Bari Leda Disum Tala Re
Bir-Buru Te Biure – Kan Gara Nari Te Singara Kan
Nimin Bugun Disum E Tain Disum Baiya Kan
Hoyo Gama Bugin Geya Joma Nuiya Lali Geya
Nimin Bugin Disum – E Ren Jonom E Kana
Anga Bira Chari Buru Dalma Chutina Singh So Te Bakai Birua Kan
Sona Gara Samon Rakha Bonga Ko Gara Biforini
Karo Kuli Marnag Gaya Te Sajao Singora Kan

[Under the forest grove, in the name of God, a country has been established named Singhbhumi. After reclaiming hills, trees and soil, by destroying the abodes of tiger and bears, by cutting the stones and reclaiming wild trees and soil we have established grounds and agricultural land. The rivers are full of water, everything is in plenty, there is no feeling of scarcity. Our country is the best of all, surrounded by four hills Angabira, Dalmachati, Chariburu are the best seats. The golden river Subarnarekha, and Koel-Karo and the big river Baitarani, the abode of Gods are here. Our country Singhbhumi is well decorated.]

On the other hand, during the hunting festival *Desh-Sikar*, a tribal song goes like this:

"Singbiree Sim Korakke
Man birre Marak Hiyoke
Cheterena regda tamark
Sade kando."

(Oh mother, the sound of cock's crow is coming from *Singbir* and the peacock's crow is coming from *Manbir*, please tell me why the *Regra Tamak's* sound is coming to day?). The mother has to reply, because to participate in the hunting festival is a must for an adolescent boy, otherwise he should be treated by his girl friends as child. The symbols of sex training and idioms of sex and the stories related to incest and taboo are told during the hunt and in the night, on the hill top of Ajodhya, the drama,' *singrai* based on sex are enacted.

To a Kudmi, another numerically very large indigenous population in Jharkhand and bordering area of West Bengal and Orissa sing a *Karam* song, which reflects the horror of coolie transportation to Assam teagardens during the colonial rule with the collaborators of the plains of India, namely with the help of Mahajan (Money lenders) and Arkathi or Middleman.

"Gai gelak bijubane
Bachchur gelak Ranebane
Bagal gelak Arunbane
Khunji Khunji-Jhamralai Mai
Tao Bagal Ghure nai aishe."

[The cows had gone to *Bijuban* and the calf to *Raneban*, and the cowboy had gone to Arunban. By searching and searching, I became really tired, yet the cowboy has not come back.]

"Hat gele Hate nai
bat gele bate nai
Bole debi he hamar saiyake
Dhudhi lote bandab Uyake."

(In the market, he is not there, in the village road he is not seen. Oh friends please tell my boyfriend/husband, I shall tie him with Dudhi-Lot, a yellow creeper. (Symbolizing the tie of love)

"Hanshi Hansi Prem Phanshi
Mahajane Paralo, Pat tular nam niye
Assamete Chalalo"

(On the pretext of love, the money-lender hanged us and to give work in tea garden they transported us to Assam). The horror of forced migration, desperation to Assam and to North Bengal, specially to Duars in semi-slavery situation is very well documented by Prof. Ashoka Sen and K. S. Singh, eminent historians of India in their various research papers.

III

Indigenous peoples of Eastern India, specially of Jangal-Mahal started resistance to the British colonialists and their collaborators against their brutal and oppressive rule, the severe draught and famines, the dominance of the upper caste Hindu elites after the permanent settlement, the emergence of the new feudal – bourgeois and their imposition of severe land taxes, construction of rail lines, transportations of cheap labours etc were the major socio-political upheavals resulting in the 1st Chuar Rebellion (1765), Kaira Vidroh (1770), Pahariya Vidroh (1789), 2nd Chuar Rebellion (1798-1800), Nayak Vidroh (1806-1816), Ho/Larkakol rebellion (1819-1920), Bhumij Revolt (1833), Ho/Kol Rebellion (1830-1832), Nil Vidroh (1830-1848), Santhal Rebellion (1855), Sepoy Mutiny (1857), Kharwar Rebellion (1872), Birsa Munda's Ulgulan (1895-1900), Medi Movement (1917-1919) etc. are the examples of people's resistance against inhuman misrule of the British colonial system. The collective identity and self-assertions of the indigenous people in Jharkhand, West Bengal and Orissa at that time revolved round the land and forest rights which they enjoyed traditionally.

One of the great culture and social activist late Mathew Areeparampil, in his book (2002), Struggle for Swaraj, very well documented the people's struggle for Jharkhand against the land-grabbers and *Thikhadars.* The Indian Forest Act (1871) and the *Thikhadars* became synonymous and thereby the indigenous and tribal people lost every right from forest which was and still life force of the livelihood. Mahto (2000) observed forests provide them Food, Fodder, Fertilizer, Farm implements, family house, building material, Fuel, Fresh air, Fresh water, Family medicine, Family or ethnic spirits and finally a place of wonderful romance. A few examples may be given here:

"*Mahul para choikre tor*
tensh phuler authre
megh kalo khosha re tor
mahokale jaire"

(Your eyes are like *Mahua* flower, your lips are like bud of the *palas* tree, your coiffure is like black cloud and having a sweet fragrance)

"*Konho Banekeri, Kiya baran ful gouri*
Chalate chalate gouri mahakale jai"

(From what forest's flowers you have ornamented your self, while walking the sweet odour is coming)

Mahato (1993) while pointing out the conflicting tradition of Ramayan between *Arya* and *Anarya* pointed out that there are several traditions.

According to H.D. Sankaliya, Ravan belonged to the Gond tribe. *Ravan banshi* Gonds are still found among the Gonds, who hold *Ravana* in high esteem as a king. To them, the word *Lanka* means a high place, an island or the home of a king. Linguistically, Gonds call all important places as 'Lakka' (not lanka). Thus according to Sankaliya, Ravan and other Rakshasas were in truth an aboriginal tribe, most probably the Gonds, and the Vanars belonged to other tribes such as the Savars and "korkus":

According to the Mundas, Santhals and Kudmis, Sita was whisked away by Ravan. The Kudmis sing the Karam songs :

"Sitake Ashok bane dekhi
Uthali uthila dudhi
Mondodari kahat ravane
Kotha pyale e ekanyadhane"

[Seeing *Sita* in the *Asoke van*, the milk of the breasts became a spring – this was narrated by *Mondodari*, who asks *Ravan*, "From where have you picked up daughter like Maiden?"] *Ravan* violated the code of incest and tribals joined with *Ram* in the battle, as they consider it a great social crime, which according to Santhal verdict *Bitlaha* (beat to death). According to Bhils of Dangs, *Savari* too was a Bhil girl of that locality. To the tribals of Tripura, *Ram* and *Sita* were the brother and sister.

IV

After independence or transfer of power, in the name of development, the total destruction of the ecology and bio-diversity started by the English educated bureaucrates, half-literate politicians and Thikhadar mafias. They never tried to understand the collective psyche and nature-man-spirit relationship. The D.V.C, the Subarnarekha multi purpose project, the Koel-Karo project and establishment of Heavy Engineering projects like Bokaro, Durgapur, Ranchi, Rourkela etc. in the Sal-forest region practically allured the job-seeker who has nothing to do with indigenous people and protection of the bio-diversity. The D.F.O. and the forest rangers and forest guards practically looted the oxygen of the indigenous people. They are the masters and the villagers living in and around the forest are their slaves or bonded labour. For last 56 years in the name of the protection of this forest the loot is given sanction by the Indian nation states. Deforestation became the rule. The various forest laws framed have only helped resource utilization than 'resource conservation'. The changed outlook together with population growth and the changing power equations between different social classes have sharpened the resource conflict of the 413 district maps of the country that are available, only 105 districts have been found to have 33 per cent of the tree cover, 52 districts have only a tree cover between 19-33 per cent, 217 district have only a tree cover of 1-19 per cent and 39 district have no discernible forest cover .

The experts opined "The living standard of Adivasis inevitably faces decline as traditional sources of firewood and forest produce become increasingly scarce. Tribal impoverishment by deforestation has to be understood in the context of tribal dependence of forests. Decrease in the collection of wood produce in tribal areas in Orissa and Madhya Pradesh has affected the nutritional pattern of the tribals. Deforestation has increased tribal indebtedness forcing them to sell much of what they collect at low price to the local merchants (Who are often the moneylender as well) to buy necessary food items. In short, today they have less access to the food they need from the forest and they do not earn enough to buy the food."

Thus we have a very recent example. The deaths of five tribals (Two Lodha and two Bhumij and one Santhal) in *Amlasole* village in West Midnapur of West Bengal are the glaring evidence. Though they live in the close vicinity of forest, they were unable to collect wood, leaves or fruits and the starvation deaths occurred. The left front government which ruled the state for 28 years has got no explanation. The ministers are also tribals, but CPI (M) leaders were forced to give explanation before the media, but no satisfactory reply was given. Similary in Ghatsila, the death of few Sabar and Birhore became an example.

Because of deforestation, hunger, poverty, starvation, malnutrition, child deaths are rampant and women are the worst victims. During my field work in 2000 and 2001 in Dhalbhum sub-division, the girl child, young and adolescent tribal women at least about 10, were sold by their parents to the outsider brothel mafias in Delhi, Allahabad, Punjab and Bombay. Because of the scarcity of food, the older generations were branded as Witch or Dahin and not less than 50 thousand tribal girls to accept the household aid job or maid servants from Assam, Tripura, West Bengal, Jharkhand and Orissa. The horrible Gonong or bride-price is now-a-days so high in

Singhbum that many tribal women had to remain unmarried. On the other hand, Muslim traders from Murshidabad and forest officials, guards, peons from Bihar and Orissa in Keonjhare and Mayurbhanj used women as their kept or rakhel.

Deforestation compels Adivasi women to walk longer distances in search of minor forest produce and firewoods vital to Adivasi economy. A study shows that Kond Tribal women put in an average 14 working hours per day as compared to 9 hours by the men. A study of mine in my own Pindra Gram Panchayat from where I became young and educated finds that not less than 76 women of the age group between 27 to 48 start their journey in early morning with BashiBhat to Purulia station to catch Howrah Chakradharpur passenger and they return back in the same train at about 8.30 of the night after collecting the forest produce and Sal leaves from the forest of Ichagarh, Chandil and also from Sini. Every day they have to pay Rs. 5 each to the Railway Protection Force and Rs. 5 to the forest guards and beautiful women are sexually abused. After returning home they prepare the leaf pot, patri and khola, and more aged women of the family remain busy to sell the materials at Purulia bazar.

The onslaught of powerful commercial and industrial interest in the plantation of Eucalyptus and Akasmani tree are rampant instead of Sal, Palas, Kusum, Kul, Mahua etc. which were once the lac producing trees, which were once considered as assets of the families. By planting those commercial trees, a beehive is never seen in those trees. Those trees are responsible for desert formation in entire Eastern India, because they suck water too much and thereby cattle and goat-sheep fodder became real scarcity. This found expression in the following songs :

"Eukilipiti gach Mahagach
Sushi lelai ho pani, sushi lelai
Sushi lelai cherikeri ghas."

(Eucalyptus trees are great trees, but it sucked the water of the grass, which goats eat)

"Pokhorer dhare, Boro Boro gach
Pathar chatan fatain dilo gachche
Pat pare mach morlo, Kahe diva kake."

(On the ponds side, the trees grow like anything even in dry stony land, but the leaves caused the death of several fishes in the pond.)

"Foresterer Bahu Gaihna Sonar
Jhabar Jhabar Kare
Duta Dalan Kolkatai, Beta-Biti Chamke
Tumkure Karntake Parhe."

(The wife of the forester decorated herself with heavy ornament of gold. He has got two buildings in Calcutta. The sons and daughters are also very smart. They read in Tumkur/Karnataka.)

Conclusion

Government of India must take a forest policy which will encourage the plantation of local and useful trees for the traditional use of the indigenous and tribal people. The Government of Jharkhand and the bureaucrates, police and administration, including Judiciary of Jharkhand state must come foreward with an open mind to give a new era to the Jharkhand people. The Government must establish a commission to study and recommend the rightful use of the forest and check deforestation. For this commission, eminent historians, anthropologists and tribal activitsts may be the members.

Text books should be prepared to bring awareness to students about the indigenous people's resistance against colonial rule so that they may feel proud. Lastly the Songs and Drama Division of Government of India. Doordarshan and Radio must take necessary steps to document and propagate our forest heritage.

REFERENCES

Areeparampil Mathew, *Struggle for Swaraj*, Tribal Research and Training Centre, Chaibasa, Jharkhand, 2002.

Bosu Mullik S., Edwin Jaydas, Anto Akkara and Anita Jayadas., *Indigenous identity-crisis and re-awakening*, Navadip Prakashan Kendra, Delhi, 1993.

Mahato, Ashok Kumar, *The Ancient – Indian Peasantry. The Kutumbins.* Read in the Asiatic Society Seminar on Transformation and Change Among the Indigenous People of Eastern India, 2002.

Mahato P.P., Ecology and Development: A Study of Folk Songs in the Tribal Peasant Area of Jharkhand, Jai Adivasi, Edited by Sarini, 1991.

______ Bharater Adivasi O Dalit Samaj, Sujan Publication, Calcutta, 1997.

______ Sanskritization Vs Nirbakization: A study of ethnic memocide and cultural violence in Jharkhand, Sujan Publication, Calcutta, 2000.

______ Tribal versions of the Ramayana, *The Telegraph – Arts*, Friday 2nd April, 1993.

______ Voice of the Adivasi/Indigenous People of India, AICFAIP, New Delhi, 2001.

Mutrowar Laxikanta, Janajati Parichiti, Adivasi Kudmi Samaj, Namkum, Ranchi, 2001.

Sen A.K., From Singhbhum to Assam; The Incidence of Labour Migration, 1873-1918, in S. Roy (ed), *History, Vol. II, No. I, 1999.*

Sen Padmaja (ed.), *Changing Tribal Life*, Concept Publication, New Delhi, 2003.

Singh C.P., *The Ho Tribe of Singhbhum*, Classical Publication Company, New Delhi, 1985.

Singh K.S., *People of India*, Anthropological Survey of India, Calcutta, 1992.

5

PHILOSOPHISING ADIVASI ECOLOGICAL PERSPECTIVE

PADMAJA SEN

Introduction

The human civilisation has as if completed the full circle, from culture springing from nature to culture maiming nature itself. As an impact of industrial revolution and globalisation we are now face to face with global warming, greenhouse gases, atmospheric pollution, pollution of soil and water, deforestation at a large scale leading to land degradation and desertification, rapid depletion of biodiversity and many more fatal consequences threatening not only the human existence alone but also any life form on this planet. The growing concern for maintaining ecological balance in an atmosphere of widespread destruction of it has inspired scholarly intervention in constructing an independent ecology-centric episteme. This paper joins the quest by formulating the problem from a different angle.

Philosophising Adivasi ecological perspective more or less contests the stereotype of Adivasi destruction of nature at large. It seeks to develop an understanding of the notion of ecology in the general socio-philosophical context and then proceeds to evaluate the ecological perspective of the Adivasis of Jharkhand with regard to forest. The paper finally delineates their ecological sensibility by enumerating their forest-related ecological ideas through their eco-centric belief system – the myth, religious ideas, festivals and an implied ecological morality towards the nature as a whole.

Ecology: The Socio-philosophical Context

Ecology is popularly understood as a branch of Biology that deals with the relations between living organisms. The term *occologie* was coined by the German Zoologist Ernst Haeckel in 1866 meaning 'the science of the relations of living organisms to the external world, their habitat, customs, energies, parasites etc.' In this sense the term implied a family of living organisms, each in close proximity to the other sharing the same physical space with conflicting appetites or complementary needs.[1] However, in practice, as Donald Worster points out, ecology has tended to focus more narrowly upon the study of 'nature', the non-human world, the world we have not in any primary sense created,[2] and to see man as an irrelevant or extraneous factor in shaping the natural environment. But that such a complete expulsion of the Homo sapiens from the sphere of nature may not be tenable very far is obvious from the fact of their reciprocal relationship. On the one hand nature/natural environment not only provides the natural habitat to human beings but also shapes their socio-economic activities, their belief system and culture as a whole. On the other hand, the homo sapiens more than any other living organism also bring about large scale changes in the natural environment. They variously distort nature. A recent study rightly observes: the awareness of man's dependence upon nature has a long ancestry; but a sense of man as the maker and unmaker of nature has only more recently dawned upon us, and with it an awesome sense of our own capacity for mischief and mayhem.[3] Hence the pure natural scientific approach to the ecological study of nature as independent of human species may not be of much relevance for the socio-philosophical study. Yet its value lies in directing our attention from extreme anthropocentrism to the autonomous natural environment.

A second lexical sense of ecology focusing on the relationship between the distribution of human groups with reference to material resources and consequent social and cultural patterns[4] is perhaps more suited for the socio-philosophical context. This approach succeeds in shifting the attention from anthropocentrism to the human engagement over time with the physical environment as an important agent and influence in shaping human history. It takes into account ecological infrastructure viz. soil, water, animal, mineral and vegetative bases of society which plays a crucial role in deciding the socio-cultural patterns. Thus in the field of social sciences this approach brings to the centre stage the ecological context vis-à-vis human agency in ascertaining the course of civilization.

A philosophical intervention into the notion of ecology however may not be satisfied with strictly issue-oriented approaches to environmental crisis. In order to arrive at more general theory relating to greater environmental issues it may try to answer such questions as – what is nature? What is the humanity's place in nature? And more importantly, what is the relationship between society and the natural world? However simple these questions may appear at the first sight, a little deliberation will make their complexities evident. For example, all of us more or less know what nature is. Genuinely the answer would be it is that which encompasses everything around us – the trees, rocks, rivers animals and so on. But then we may ask, is not a garden carefully grown, nature? Is not the house with all its furnishings a part of nature? And for that matter, are the human beings and the society less part of nature than animals? Such questions finally lead us to the ethical import underlying environmental issues. If human beings are part of nature they are not merely one among the many living forms. They are unique in the sense that they are moral agents having responsibilities towards the ecological infrastructure, which no other species is capable of sharing.

A philosophy of ecology thus addresses itself to an ecology-oriented ethical sensibility that may provide a standard for human conduct towards the nature as a whole because human beings have duties to eco-systems. The environmental ethics, a comparatively new branch of philosophy, is in this sense the 'attempt to expand the moral framework to nature and counter human chauvinism by showing that feathers, fur, species membership and even inorganic composition are not barriers to the range of ethical considerations.'[5] The environmental ethicist Paul Taylor defends the adoption of a bio-centric ethical attitude of respect to nature. He grounds this attitude in the intelligibility of regarding each living entity as striving to realise its own good and as having the same inherent worth within a network of teleological centres of life.[6] Thus to act with an ecologically creative ethical sensibility is to act responsibly and morally towards the nature at large.

The Eco-centric Belief System of the Adivasis of Jharkhand

The long association with the region, its forest and nature at large induces Adivasis of Jharkhand to evolve an eco-centric belief system. Living in close proximity with nature has made them feel to be part and parcel of it. Hence they have developed a bio-centric ethical attitude of respect for the natural objects – the mountains, rocks, streams, forest, plants and animals assuming that they have an equal inherent worth within the natural network. It is implied in their myths, religious ideas and rituals, festivals and social institutions.

The Myths

The creation myths of the Adivasis relate in a very clear manner how the earth, plants, man and animals are created by *Haram* or *Singbonga*, the Highest God, as well as how the symbiotic relation between the spirits, man and nature at large is preserved. According to a Ho myth the earth was created first, and then it was clothed with grass, tree, rock and water. Then followed cattle and all the wild animals. At last *Singbonga* made a little boy and a little girl. The Munda myth opens the narration: 'When the sky was empty and earth all covered with water, the Old One had made only animals that lived in water.' He then sent crab, tortoise and earthworm in turn to collect some earth from below the water. Of these three only the last could bring some earth. *Haram* multiplied this to make dry land. On this he made all kinds of herbs, plants, trees and animals. He then made the first clay model of man, which was kicked by the horse. So He made another figure again and enlivened it. Another version of creation myth relates that after the creation of dry land and animals a giant stork laid two eggs, hatched them and brought forth a boy and a girl.[8]

The Santhal myth also narrates some such story with a little variation in the order of creation. It states that first *Thakur-Jiu* had created the beings, which lived in water. Thereupon He made two clay models but before He could give them souls (life) the day-horse came from above, trampled them to pieces and left. *Thakur-Jiu* then decided to make birds but this time not of earth. He made *Has* and *Hasil* (a pair of swan) pulling (material) from His breast. He breathed on them. They became alive. *Thakur* then created dry land. He caused the earth to be harrowed level. Mountains, all the vegetation, plants and trees were also gradually created. Two birds then laid two eggs from which two human beings were born, one boy and one girl.[9]

Some scholars may sneer at the apparent fictitious and illogical character of the above myths. But I am not concerned here with a discussion as to the reality of the myths extended by different schools of anthropologists such as Levi Strauss, Durkheim and Marcel vs. Malinowski, Eliade and others. Symbolic in character as the myths are, their truth lies in the idea they seek to convey. The myths express the Adivasi belief that Homo sapiens are the product of the same creative process as the other objects of the nature are, and hence they are one among them, the part and parcel of the greater nature.

The Religious Beliefs

The religious beliefs of the Adivasis also exhibit their close link with nature. Besides *Singbonga*, the belief in a pantheon characterises their religion. The pantheon includes a host of nature gods who frequently appear in their myth. According to a Ho myth, being offended with creation, *Singbonga* once released blazing flames or rain of fire, which destroyed all the living creatures on earth. But *Nage-era* (river goddess) saved two souls, a brother and a sister under a vast sheet of water. Other than *Nage-era*, *Maran Buru* (mountain god), *Jahera-era* (the lady of the sacred grove) and the Asur women whom *Singbonga* had turned into the *Bongas* of nature also appear in myth.

The Adivasis do not believe in erecting temples, churches or mosques as the places of worship. They hold natural objects like forests, hills and rivers to be the best abode of gods and spirits.

Hence nature is considered as sacred. This feature of their religion characterises them as naturalistic as opposed to the institutionalised religions like Hinduism, Christianity and Islam. In consonance with this idea, traditionally every tribal village has a sacred grove consisting of a part of original forest and *sal* trees, which is named as *Sarna*.[10] It appears that like many forest-dwelling traditional societies,[11] Jharkhand Adivasis also evolved the concept of sacred grove apparently due to religious belief, for fear and respect for god and ancestral spirits. It served as the place of worship and performance of religious rituals as well as socio-cultural and religious interaction. However it is not very unlikely that the indigenes, with their long association with nature, would have known with experience the utilitarian value of bio-diversity contained in such groves and would have insisted on nurturing one in the close proximity to the villages for all practical purposes. Belief in sacred totem also helped in maintaining bio-diversity. But unfortunately with time this traditional wisdom is losing its importance.

Festivals

The festivals of the Adivasis closely follow the natural and annual agricultural cycles. Here I would mention only a few familiar and common festivals of Kolhan in West Singhbhum to illustrate the cultural link of the Adivasis with nature.[12] These festivals are observed to invoke the blessings of Gods and spirits to increase the fertility of the fields, to protect the crops from natural and supernatural calamities and for thanksgiving to the village and tutelary *Bongas* for the well-being of the tribe.

The *Maghe parab* consisting of five days is observed in January-February. In importance it corresponds to the *Bandhan* or *Sohrai* festival of the Santhals. The festival is meant to offer sacrifices to *Desauli Bonga* and village tutelary god to save from epidemics and to bring down rain. The flowers of *palas* tree are offered to God.

Baha is the festival of flowers. It is held in the months of March-April in the honour of the village *Desauli*. It corresponds to the *Sarhul* festivals of the Mundas. The offerings of new *sal* (*sarjôm*) flowers are made. People also decorate themselves with *sal* flowers. The custom symbolises the intimate relationship of Adivasis with the *sal* forest of the region. One cannot eat new fruits of the jungles or use *sal* leaves for making leaf plates and cups until *Baha parab* is over.

Damurai festival is celebrated in Kolhan in May and is related with the sowing of paddy. *Hero parab* is observed in June at the time of transplantation. Prayer is offered to *Desauli* and *Jahira buri* to receive their blessings for good harvests. *Bahtauli* is performed for protecting the paddy crop and its maturity.

Moreover, the festival of *Jomnam* is celebrated in August-September when *gora* crop ripens. First fruit of the harvest is offered to *Singbonga.* Till the sacrifice the new rice is not eaten. *Kalam puja* is observed in October at the removal of the paddy straw from the *Kalam* (threshing floor) to be staked.

Philosophy of Nature

Thus an eco-centric belief system along with sustained dependence upon nature induced the indigenes of Jharkhand to evolve a distinct philosophy of nature. Their creation myths imply that different elements in creation have a symbiotic relation among each other based on a sense of equality and mutuality. Hence, the smallest of the small earthworm and the big one like a tiger, all collaborate in the process of creation.[13] *Haram* commands the animals to serve man but he also assigns the man the duty to look after them. In the myth relating to the division of time,[14] *Haram* explicitly commands man to regulate his life according to nature. The existence of man is dependent on nature as he draws all succour from it but it does not empower him to have a command over nature. The Asur legend of the indigenes states that though *Singbonga* punished the Asurs, he compensated their wives. He made them *Bongas* or spirits of nature subsisting on the sacrifices made to them by the Adivasis.[15] Thus a symbiotic inter relation is established between man, nature and spirits. This is why the *Adivasis* continue to consider perfect harmony and peace with the community, nature at large and the supernatural powers as the essential element of their ethnic value system.[16]

Forest and the Adivasi Ecological Perspective

With this socio-philosophical and religious context in mind it would be easier to evaluate the forest related ecological perspective and the role of Adivasis of Jharkhand in it. Adivasis are perceived as the so-called destroyer of forest, a view generally held since British rule.[17] The fact cannot be denied that with the growing pressure of population a great deal of original forest has been denuded by them for village making and cultivation since before the on set of British rule. Yet compared to the state-sponsored destruction of forest for mining, industries, railways and sometime for sheer profit making, clearing of forest by the Adivasis must be very meagre. The actual difference may perhaps be adduced from the official records. Moreover whatever destruction the latter had caused has been done for existential purposes. No traditional Adivasi household exhibits the misuse of wood for luxurious furnishings. On the contrary a strong and continuous link with forest characterises Adivasi life and culture. Forest for them is not merely an inanimate large uncultivated tract of land covered with trees and under wood or pasture. It has been a shelter and basis for survival in the moment of existentialist crisis, which they have had been habituated with a sustainable use. How forest has been instrumental in shaping their culture is evident from their eco-centric belief system related above.

What I want to emphasize in this section is that, if viewed from a standpoint of technologically equipped social groups, the Adivasis seem to be in no position to exploit the forest of their area beyond a limit. During the stage of hunting and food gathering and also shifting cultivation they were dependent on forest for mere survival. Supplies from nature being more than their collective need they were a happy and contented lot. Absence of technological knowledge and dependence mainly on human muscle power also prevented them from misusing the forest beyond a certain limit. Their knowledge base being fairly low they were in no position to control the nature. Nature appeared to them capricious and mysterious. This evoked in them a deep sense of fear and reverence for it.

Naturally in this early phase their ethnic value system inspired the indigenes to assign sacredness to the objects of nature, such as mountains (*Maran Buru*), rivers (*Nage-era*) trees (*Sarana*, the sacred grove) etc. of which they considered themselves to be a part. This positive felt relationship coupled with deep sense of psycho-religious fear and reverence towards nature induced them to use natural resources particularly forest products only according to their meagre existential needs and demands on a sustainable basis. A detailed future study however is needed to underline their precise mode of use of natural resources in general and/or forest and forest products in particular.

Even at the present stage of socio-economy the Adivasis of Jharkhand forge a very special relationship with the forest. They like themselves to be known as forest people, the root of which lies primarily in their historic dependence on forest for survival. They still have not been able to shed off the memory of their wildness and their original link of the hunter-gatherer stage with the primeval forest. Hence they feel a natural right over the forest and forest products. A recent study on Ho differential peasantisation remarks that it had a basic ethnic character about it, constituted as it was by an essential yet variable link with forest. Agriculture being largely dependent on monsoon, varying use of forest and forest products served as a supplementary base for the Adivasis.[18] Quite naturally infringement into their natural rights triggered social protest right form British period ranging down to the present.[19]

However with time, modern education and use of existing technology the indigenes are no more solely dependent on forest. With experience and long interaction with forest they have also accumulated the required knowledge as to the utility of the bio-diversity of forest for their survival. On the basis of their traditional ecological knowledge they have evolved such socio-cultural practices as to develop *Sarna* to actively maintain the bio-diversity and productivity of the forest for their use. The general symbiotic approach of the indigenes towards forest seems to originate from an attitude of respect for it. Their tradition and customs endorse a use of natural resources with constraint and on substantial basis. Such approach implies a much-valued eco-centric moral attitude towards nature.

Conclusion

We find a drastic change in value-perspective if we compare the survivalist compulsion of the Adivasis coupled with their eco-centric moral attitude towards nature with the modern consumerist approach to natural resources, particularly in respect of forest. In about two centuries the ecological impact of Industrial Revolution has been profound and far surpassing all other preceding epochs of human civilisation. Away from nature, armed with the latest technology and scientific knowledge and accompanied by a consumerist attitude to life, man no more felt himself as a part and parcel of the sacred nature. He believed himself to be the master of it, having every right to exploit it for his own well-being and luxury. He felt no need to be prudent towards natural resources. The result has been devastating. Two paradoxes with respect to forest use illustrated by a recent study are eye openers. First, more the spatial separation from the forest, the greater the impact on its ecology, and the further removed are the actors from the consequences of this impact. Second, the faster the development of formal scientific knowledge about the composition and functioning of forest type, the faster is the rate of deforestation.[20] Unfortunately the worst sufferers of such unwise acts have been the forest people. A blind faith in the capability of scientific knowledge and technological changes for providing an infallible guide to the resource use has all the more encouraged major interventions in natural eco-systems. The history of forest management is replete with illustrations of the failure of sustained-yield methods to forestall ecological collapse.[21]

Thus, ecology should not be viewed as autonomous of the Homo sapiens. More they assume the role of arbiter of nature the more will be the relevance of a linkage between ecology, social sciences and the value system for building bridges with nature. Religious beliefs and customs of traditional societies can play an important role in achieving this goal. But unfortunately in this era of information technology and global village they are also fast loosing their traditional wisdom.

NOTES AND REFERENCES

1. Cited in D. Arnold & R. Guha (ed), *Nature, Culture, Imperialism. Essays on the Environmental History of South Asia,* Oxford University Press, New Delhi, 1995, p. 2.
2. Donald Worster, 'Doing Environmental History', in Donald Worster (ed), *The Ends of the Earth: Perspective on Modern Environmental History*, Cambridge, 1998, p. 292.
3. Arnold & Guha, p. 3.
4. Webster's Dictionary, Encyclopedic Edition, J. G. Ferguson Publishing Company, Chicago, 1979, p. 574.
5. T. Honderich (ed), *The Oxford Companion to Philosophy*, Oxford University Press, New York, 1995, p. 238.
6. *Ibid.*
7. Lt. S. R. Tickell, 'Memoir on the Ho Desum' (improperly called Kolehan)', *Journal of Asiatic Society of Bengal*, Vol. XI, Part II, 1840, p. 797.
8. J. Hoffmann, *Encyclopaedia Mundarica*, Vol.13, Gyan Publishing House, New Delhi, Reprint 1998, pp. 3981-88.
9. P. O. Bodding (Tr.), *Traditions and Institutions of the Santhals/Horkoren Mare Hapramko Raek Katha*, Bahumukhi Prakashan, New Delhi, Indian Reprint, 1994, pp. 3-7.
10. P. Sen & A. K. Sen, 'Religion as Identity: Evolution of Religious Ideas among the Adivasis of Jharkhand' Special Issue: Philosophical Traditions in the Indian Languages, *Journal of Indian Council of Philosophical Research*, 2002, pp. 247-58.
11. One such grove is the *Mawsmisacred grove* maintained by the Khasi tribe in the Cherrapunji region of Meghalaya. P. S. Ramkrishnan, *Ecology and Sustainable Development*, National Book Trust, India, New Delhi, 2001, p. 41.
12. The other Adivasi groups also have corresponding festivals with a little variation. But to mention all of them is not possible here.
13. Hoffmann, *Encyclopaedia,* pp. 3981-83.
14. *Ibid.,* pp. 3985-88.
15. *Ibid.,* pp. 240-42.
16. P. Sen, 'Value, Valuation and Tribal Society', in P. Sen (ed.), *Changing Tribal Life : A Socio-Philosophical Perspective*, Concept Publishing Company, New Delhi, 2003, p. 118.
17. A.K. Sen, 'Protected Forest Rules and the Ho Social Protest' in this volume.
18. A.K. Sen, 'Differential Peasantisation and the Changing Socio-Economy of the Ho Tribe of Singhbhum', presented at the Seminar on the Changing

Scenario of the Tribals and Indigenous People of Jharkhand, Bihar, West Bengal & Orissa, at *The Asiatic Society*, Kolkata on 8 January 2003.

19. See papers by A. K. Sen and C. K. Paty in this volume.
20. M. Gadgil & R. Guha, *This Fissured Land*, Oxford University Press, New Delhi, 1999, p. 52.
21. *Ibid.*, pp. 52-3.

6

CHANGING ECOLOGY AND ITS IMPACT ON LODHA TRIBE OF WEST BENGAL

AMIT SAURABH PATY

Introduction

In the Indian map, the tribal peoples have lived within or adjacent to the forest areas. Their entire life has been rotating round the forest. So they had made an intimate relationship with the forest ecology for their survival. But, sometimes, we can see the indefinite and radical changes in the relationship between community and forest ecology due to various reasons. Moreover this causes a dangerous effect for few communities as it happened with Lodhas who are one of the most primitive and backward tribes of West Bengal. It is the history of this nomadic Lodha tribe who were once displaced from their original habitat, which began more or less with the advent of colonial rule and due to this ecological change, they went ahead to the path of criminality. Now they are known also for their criminal activities, though they have taken to a life of crime for other reasons. Thus, the paper in brief explain how a simple forest-dwelling community–Lodha- turned into a criminal tribe and also what are the gradual impacts on their life-style, in a historical perspective.

In the state of West Bengal, the Lodhas are anthropologically not only more interesting but they had also played the most notable role in the history of forest dwelling communities of this region. At present, they mainly reside in Midnapur district in a multi-ethnical environment with Santhals, Kheria, Oraon, Bhumij, Mahlis, Mundas, Mahtos, Brahmans and Sadgops.

Besides Midnapur, they also live in Mayurbhanj district of Orissa and east Singhbhum district of Jharkhand.

The word *Lodha* might have originated from the word *"Lubdhaka"* meaning a *trapper* or *fowler.* Similarly, another theory is that the Lodhas came from Ludhiyana district of the Punjab. They adopted the appellation of Lodha or Ludhi after the name of the place of their origin. The other is that the Lodhas had adopted the name from the *Lodh* tree which grows in abundance in Northern India[1]. But due to the lack of detailed information, it is really difficult to ascertain the exact nomenclature of the tribe.

Due to this, different colonial-scholar administrators described them in various ways. Risley considered them a sub-tribe of Bhumij of Chhotanagpur[2] while Hunter identified them as a Hindu caste[3]. This confusion in identifying Lodhas continued during post-colonial days. In the census of 1951, they were considered as a caste along with Savars.[4] But the census of 1961, tentatively classified them under the Austric family of languages. In the census of 1971, the Lodhas were merged with *Savar* and *Kheria* tribe to dilute their independent identity. Now, they were termed as *Lodha-Savar* and *Kheria Savar*[5]. Meanwhile since 1956, the Lodhas came to be recognised as scheduled tribe, and later they were categorized as primitive tribal group in 1982, a special weaker section among the tribal in need of special care by the government of India. Lodha population was 53717 according to 1981 census[6]. They followed a typical patriarchal family and were grouped in a few exogamous divisions.

Over centuries, the land and forest had always played an outstanding role to provide protection and sources of livelihood of the many tribal communities living in this region. Akin to many forest- dwellers, Lodhas also lived mainly in the jungle of Midnapur district and depended on hunting, and collection of edible minor forest produces like roots, nuts, tubers etc. From many years, in rural Bengal the collection and trading of jungle products were mainly existed by Bhumijs, Santhals, Lodhas and Manjhis[7].

The pattern of their general economic life reveals that they were involved in a few subsidiary occupations and their major professions are directly linked with the economy of their surrounding ecology. From the jungle they catch snakes and sell the skin to merchants. Collection of firewood and mat making is an important economic pursuit of Lodhas, specially the aged persons of the community[8]. Fish catching during the rains or autumn and selling it to their prosperous neighbours is another occasional vocation of Lodhas. They used to grow tusser cocoons from the *Asan* (Terminolia Tomentosa) trees and collect *mahua* flowers or fruits to supplement their income.[9] Besides this they used wild fruits, roots, and flesh of small animals as a content of their meal. Thus their food habit and economic activities were very closely interwoven with their habitat.

A good number of clan names of the Lodhas are associated and identified with the forest resources and animals such as *mallik (*Sal fish- a kind of fish), *dandpath* (tiger), *pramanik* (a kind of bird) etc. Their God/Goddess, folktales, myths and songs reflect their close relationship with forest. The *Sal* and *Mahua* trees are considered as the seat of their village deity *Gram* or *Baram*. *Joychandi* or *chandi* believed to reside in the jungle, is the deity of Lodhas[10]. Thus Lodha's socio-economic life, their culture, and traditions were always interrelated with forest. This was the earlier pattern of their tribal life. Nevertheless this identity remained intact for centuries as long as the forests around them were intact.

Obscure history

The previous history also tells that there was a symbiotic relationship between the Lodhas and the forest and they used forest resources without stress and strain. They had almost depended and resided in the deep forest. They roamed freely in the forest area of Midnapur. Most of them were satisfied with the collective forest economy of hunting and collection of forest produces with the freedom of wandering in the forest. But the introduction of 'Permanent settlement and other land related measures' framed by the East India Company for the revenue and other purposes curtailed their erstwhile freedom of living. It was applied to many districts of Bengal as well as in Midnapur district. Consequently the traders, merchants, local suppliers and contractors became modest and small zamindars or landlords.

At that time Midnapur comprised of 57 parganas, out of which 42 parganas had been given to these Zamindars. In this way the social mobility process allowed some castes, tribes and other communities of rural Bengal like Sadgops, Bhumij, Tellis or Kshatriyas to become dominant groups as landlords in that region. By this many isolated forest tracts and other *Khas* lands were permanently leased out to these Zamindars[11]. This unjust land settlement moreover gave non-tribals control over vast areas of isolated jungle tracts of Lodha habitat who depended entirely on forest for their sustenance. Naturally, Lodhas had been pushed back from their old hunting grounds.

At the same time, the forest in which the Lodhas lived had been encroached upon by a few agricultural communities like Santhals, Mahtos, and Sadgops. They cleared it for the cultivation and different other purposes[12]. It was some kind of gradual interference in the natural habitat of Lodha. This happened because the Lodhas were not as strong as the other groups. At the same time, they had also taken advantage of the fact that Lodhas were ignorant of agricultural operations and were still interested to follow the same archaic forest economy of collection of jungle produce and hunting and trapping of animals. The situation worsened because the government was also on the side of cultivated classes, who paid revenue to the government. Thus they were prohibited from free access into the forest and its use for their different existential needs. In this way they were deprived of their hold and dependence on the forest.

The condition of the forest dependent Lodhas worsened also because of the progressive dwindling of the forest around them. Consequently, reckless cutting down of the forest became the order of the day. Hardly adequate forest was left for them to collect and to forge. It displaced them from their eco-system and forced them to migrate eastern part of Midnapur district which was less forested and had plenty of plain land, which was useless for them[13].

Unlike other neighbouring tribes, the Lodhas failed to overcome these tricky situations which affected their traditional form of economic life very seriously. Obviously, they were exposed to a new environment with new socio-economic forces. Continually they were facing the unsympathetic attitudes and oppressive incidents from government and other communities. All these unpleasant incidents made them ever more disorganized, cynical and anti-social. Probably under these scared circumstances, hunger and needs of other essential belongings forced a number of them to go against the society. Hence in the struggle for survival they preferred to be involved and engaged in robbery, theft and other criminal activities. They might consider this path a convenient means of livelihood. Possibly it was first started individually but, in course of time, it became the practice of several groups of Lodhas like Doma and Bhalkishol gang[14] who were later recognized for committing a good number of systematic and organized crime. Thus amidst poverty, unsympathetic attitude of the neighbours and stoic apathy by the ruling government, criminality cut a deep gorge into their society in which these people had to roll down helplessly[15].

The Lodha Categorized as a Criminal Tribe

The British government was facing many administrative difficulties and problems like murder, dacoity etc. in many places. It was a feeling of government that these offences were generally committed by some communities who were hereditarily criminal by profession[16]. This could only be controlled by some strict law. So the British government passed the Criminal Tribes Act in 1871 to control the criminal activities of such groups. An amendment of this act in 1897 empowered the local admini-stration even to segregate the children of those groups aged between (04-18 years) from their parents. The act made such offences non-bailable.

Another amendment of 1911 made it compulsory for the members of these tribes to give their fingerprints and to report the change of address to the nearest police station. It authorised the government to restrict their movement within particular areas. In whole it placed certain communities under strict supervision. The background of the Lodhas being brought under the purview of the act was as follows. In 1900 there were many criminal cases like theft, burglary, dacoity etc. committed by Lodhas. It created problems to the administration. So eventually Lodhas were categorized and declared a criminal tribe in 1916 under Government Notification no.7022-23 of 20th May, 1916[17].

Up to 1937 several amendments made the Act increasingly more stringent to augment the sufferings of the Lodha.

Socialization Process

Nevertheless in 1937, the Criminal Tribes Committee[18] headed by V. N. Tiwari observed that the criminal activities of these tribes were the result of the harmful social atmosphere. So it recommended to repeal this Act. But only in 1952 the Criminal Tribes Act was repealed. Hence, about 2,71,02,180 people of 153 communities, including the Lodhas were freed from the stigma of criminality[19]. At this time, with the formation of backward commission, Government of India made suggestions as they should be called 'De-notified *Vimukta Janjati* (ex-criminal*)*, and should be included in the category of either Scheduled Tribe/Scheduled Cast or Other Backward Community. At the same time it also decided that rehabilitation should be provided to them.

Immediately after the revocation of Criminal Tribe Act the initial measures were taken by the Home (Police) Department covering all the Lodha-centred villages in the district of Midnapur. Some administrators and few social workers arranged regular meetings and discussions for creating awareness among the people about the issue of Lodha's socialization and rehabilitation process. It was much appreciated by the Lodhas themselves because after this move the torture and harassment by police department were immediately stopped. Soon the government started many schemes and projects among the Lodhas. Some voluntary organizations also started welfare schemes by receiving financial assistance from government. Moreover a number of multipurpose cooperatives and lampus were formed. Some lands as well as agricultural implements, seeds and bullocks were provided to transform them into a settled agrarian community. At the same time various subsidies like goats or cows were made available to them.

In this rehabilitation process among Lodhas, Dr. P.K. Bhowmick, an anthropologist, played a significant role. He rightly considered that their criminality had been due to the disintegration of their social moorings, and their economic and territorial displacements from time to time. To raise these unfortunate people from the abysmal depths of ignorance and backwardness, he formed many self-help groups, village committees, vocational and educational schools and hostels[20]. The above efforts helped the Lodhas to be changed from the hunting-gathering people (or/of criminality) to agricultural ones. The intention of all these efforts was to provide them a stable economic background as well as to introduce cultivation among them to bring them at par with all other progressive communities of the region.

Action and Reaction

The honest intention and effort made to rehabilitate the Lodha to a life of dignity and self-respect did not however fully succeed. The greater society more or less failed to rationalize the changed image of the tribe and continued to suspect them for their past activities[21]. It proved the saying that a "lost good name is never retrieved"[22]. Furthermore, the non-Lodha communities became jealous of Lodhas to some extent for the care and assistance that they received from the on-going schemes. Out of reaction other communities started to indulge in killing of Lodha people. Though Lodha- killing had been a regular feature in West Bengal but since 1961 a spurt in it was visible. In that year a mass attack was made on the Lodhas by other tribal groups. This was repeated in 1968, 79, 81, 82, 84, 86, and 1989.[23] The general trend was to impute the blame on the Lodhas for any or every crime committed in the locality and to indulge in their killing.

Mahasweta Devi, a well-known human right activist, who works with Lodhas in West Bengal has presented a brief analysis of the situation. She has observed that though Lodhas were de-notified in the 1950, the rural population still thought of them in the same way. Whenever there is theft or dacoity, it is customary for the police to harass the Lodhas. *Although it is well known to the police and the administration that the upper class village worthies are the brain behind these efforts, the Lodhas being only their tools and the receiver of the stolen goods are the real criminals. But no one dares to touch them*[24]."

At present other communities also admit that in comparison to the past their criminal activity is less than 2-3 per cent instead of 90 per cent in the locality. But a deep prejudice prevails among the neighbours, starting from the children up to the educated people, the social workers and even those social scientists who work directly with Lodhas. They are still presenting the Lodha first of all as an ex-criminal tribe, instead of a scheduled tribe or simply a primitive group[25]. The burning example was the case of Chuni Kotal— a 27-year-old graduate girl from the Lodha community. The stigma of criminal tribe is so deep rooted that she has always been harassed by the society. The harassment forced Chuni to commit suicide on 16 Dec. 1982[26]. This single incident may be an eye opener and sufficient for explaining their social negligence and exploitation by the civilized world for centuries.

Moreover the Lodhas still find it very difficult to adapt themselves to an unusual system which was so different from the original one when they lived in forest. It has been found that they prefer working as day labour rather to cultivate their own land because cultivation required time and patience which they lacked. So what happens is that they sell their own land in lieu of money or paddy. Similarly they failed to maintain the cow, goat and bullock which had been offered to them by the government. Often, they tended to sell their cattle to others[27]. Actually due to their traditional link with the forest ecology they still find a forest based economy rather than agriculture conducive to them.

It is a welcome development that, in recent years, the government is trying to provide and restore their original environment through reforestation programme in some selected areas[28] but there are very few who are enjoying this ecology because it is also now irrelevant to them because of the rupture that had been already convenient to their life-style. Over five decades, a huge amount has been spent in the name of Lodhas but we find that the community has still low literacy, instability and live a life of acute poverty.

Conclusion

Thus it becomes clear that a tribe rooted to a life-style and culture built around forest gets disrhythmed and distorted when they are evicted from their natural setting. How so ever honest be the subsequent effort of socialization the tribe unfortunately fails to acclimatise itself to the new setting and way of life. The problem is compounded when the greater society fails to dispel the prejudice against the community which creates an atmosphere of disbelief. Restoration back to the past cultural setting also meets with sad setback because of the rupture created meanwhile. One may ultimately be forced to conclude that displacement of the forest people from their ecology for earning greater revenue or to privilege the affluent and mighty at the cost of the poor tribals ultimately marginalises the poor and benighted whose rehabilitation turns out to be an uphill and mostly a failed venture.

NOTES & REFERENCES

1. P. K. Bhowmick, *The Lodhas of West Bengal – A Socio-Economic Study*, Punthi Pustak, Calcutta, 1969, pp.10-14.
2. H. H. Risley, *The Tribes and Castes of West Bengal*, Vol. 02, Bengal Secretariat Press, Calcutta, 1891, p. 21.
3. W. W. Hunter, *A Statistical Account of Bengal*, Vol. III, Trubner & Co. London, 1876, p. 50.
4. P. K. Bhowmick, op. cit., p. 280, Many times, they themselves also consider savar which is actually an ancient snake-charming & mythical forest dwelling community of legendary repute in West Bengal.
5. Mahasweta Devi, *The Lodhas of West Bengal*, in Maitry Ghatak (ed.) *Dust on the Road: The Activist Writing of Mahasweta Devi*, Seagull Publication Private Limited, New Delhi, 1997, p.114.
6. R. C. Sharma, *Indian Tribes Through the Ages*, Publications Division, Ministry of Information and Broadcasting, Government of India, New Delhi, 1995, p. 42.
7. Ranbir Samadar, *Memory, Identity and Power: Politics in Jungle Mahal-West Bengal*, (1899- 1950), Sangam Publication, New Delhi, 1998, p. 37.
8. Amit Saurabh Paty, *Role and Status of Lodha – aged of West Bengal* – M. A. Dissertation, Department of Anthropology, Post Graduate Department, Ranchi University, 2003, p.15.
9. P. K. Bhowmick, *Applied Action Development Anthropology*, Institute of Social Research and Applied Anthropology (ISRAA), Calcutta, 1990, p. 207.
10. P. K. Bhowmick, *The Lodhas*, op.cit., pp. 160.
11. Hitesranjan Sanyal, *Social Mobility Process in Bengal, The Indian Historical Review*, Vol. II, No. 01, Vikas Publication House Private Ltd., New Delhi, 1975, p.79; also See N. K. Singh, *The Economic History of Bengal from Palassy to the Permanent Settlement*, Vol II, Sri Gouranga Press Private Ltd, Calcutta, 1962, pp.70-83.

12. P. K. Bhowmick, Applied Action, op.cit., pp. 201-207.
13. P. K. Bhowmick, *Problems of Denotified Tribes: A Case Study of the Interaction of Government and Diverse Ethnic Groups in Fringe Bengal*, Man in Life, Vol I & II, New Shakti Press, 1976, Calcutta, p.13.
14. P. K. Bhowmick, The Lodhas, op. cit., pp. 271-273. This criminal group was a Lodha-constituted gang, covering a wide area of Midnapur, Jhargram and Kharagpur Police stations. It has committed altogether 91 dacoities, 24 robberies, 35 burglaries, 37 thefts up to the end of the year 1953, pp. 271-273.
15. *Ibid*, p. 280.
16. R. K. Gupta, *The Working of an Act: The Case Studies of the Lodhas and the Karwal Nats since 1952*, Man in life, Vol. I & II, New Shakti Press, Calcutta, 1976, p.71. He illustrates the British Government's attitude towards the Lodhas in this paper. For this, he referred to the remark of Mr. T. V. Stephen–a member for law and order in the Viceroy's Executive Council in 1871. The Committee has related the criminal activities of a certain group by heredity and sought to suppress these born criminals by stringent laws.
17. P. K. Bhowmick, The Lodhas, op.cit., p.160.
18. P. K. Bhowmick, Problems of, op.cit., p. 13. It was a committee of socio-legal nature, headed by Mr. V. N. Tiwari to enquire into the working of the Act and to offer recommendation for its modification or repeal.
19. R. K. Gupta, op.cit., p. 72.
20. P. K. Bhowmick, Applied Action, op.cit., p.63.
21. Sarita Bhowmick, *Lodha Odyssey: A Journey between two Worlds*, Man and life, Vol. 08, No. III & IV, Samaj Sevak Sangha Press, Calcutta, 1982, pp. 172-174. She describes in her paper, the combined physical, mental, and social profile of Lodhas life. They were in general, brutish, selfish and very expert thieves. They could cheat or deceive anyone and all sorts of misdeeds could be done in the locality by a Lodha. So, for centuries, a very low image about the Lodhas, among the others, developed in course of time.
22. Paulo Francesco De Leo, *Bidisa: Role of Voluntary Organisation in Development of Lodha Tribe,* M.A. Dissertation, Department of Anthropology, Post Graduation Department, Ranchi University, 2003, p.54. In spite of all these good efforts, unfortunately even if the theory elide, the mark of criminality remained and recalled by the new denomination and since now the Lodhas are still presented as "ex-criminal."
23. Sarita Bhowmick, op.cit., p.173.
24. Mahasweta Devi, op.cit., p.116.
25. Paulo, op.cit., p.29.
26. Mahasweta Devi, op.cit., p.139; also see in Mahasweta Devi, Adivasiyon ke Beech in Jharkhand perisist-2, Prabhat Khabar, Neutral Publication, Ranchi, 2002, p. 4.
27. P. K. Bhowmick, Problem of Denotified, op.cit., p.17.

28. Dipur-an originally Lodha village in Narayangarh Police Station of Midnapur district where reforestation process is going on by government but still it has not got success.

7

ECOLOGICAL IDEOLOGIES IN TRIBAL FOLKLORE OF EASTERN INDIA

ADIKANDA MAHANTA

From the historical research it is revealed that the life style of the tribals of eastern India is circumferenced by not the coastal belts, but by the hilly areas. The structure of life of these tribes and the indigenous people is closely linked with forest, and its flora and fauna. These inhabitants i.e. Santhal, Munda, Kudumi, Dharua, Ganda, Bhuyan, Bhumija, Birhor, Lodha, Khadia, Ho, Mahali, Mankidia, Kolha, Orang, Bathudi, Puran etc, have their own identity because of their individual (own) communal culture.

Some non-tribals from outer regions who have not come in contact with these tribals take it granted that the tribal means a caste or a community. These non-tribals have a wrong conception that similar types of norms and social traditions flourish among all these tribes. Their social life is steered by a stereotype and monotonous social taboo and belief. Their folk literature is homogeneous and different from that of the non-tribals.

But all who have associated with the life style of the tribals know that tribal does not mean one caste or one community. Moreover it is a conglomeration of many tribes or an amalgamation of many communities. In addition, the language, norms of conduct, belief, thought, social taboo and life style are different from one tribe to the other.

An example will better explicit it. In eastern India, Santhal, Bhumija belong to one category. The two groups inhabit in close vicinity. But their social structure and taboo of life is different. There is provision of the immersion of bones of a Santhal who is dead in the water of a holy river. It is entitled as "Damodar yatra" but the Bhumijas bury the bones of the dead. It has become a ritual of this community to cover a stone over the buried corpse.

Similarly, if we study the system of marriage. The Santhal communities believe in the conversion of the *gotra* of the bride after the marriage. It means the bride of one *gotra* of this community is identified by the *gotra* of the groom if she marries. For instance, if the bride of *Tudu Gotra* is wed-locked with the groom bearing the Gotra of Hembram, she is included in *Hembram gotra.* But the Bhumijas are not in favour of this belief. In the norms of the community, the bride is identified by the *gotra* of her father. There is a ritual of the cremation of the bride after her death in the name of the *gotra* of her father. There also exists the ritual to bury even her bone after bringing it to any cremation ground of father's *gotra* if it was not possible to bring the whole body in any unavoidable circumstance.

In this way, there are differences in taboo, beliefs and norms etc. between each tribal and the indigenous people. It is not only applicable in the field of this type of rituals, but also in the field of language. While some communities speak 'Austric' language, the other speak in 'Dravidian' language and the language of some other tribals belongs to 'Aryan' language.

In this context, one may be doubtful about the fact that the philosophy of all the tribals inhabiting in eastern region rightly reflected in their folklore while discussing the ecological ideology. But this writer believes that all the tribes and the indigenous people have the common origin, i.e, forest to each other. Therefore, strange tunes and rhythms of drums are heard in different festivals with lucidity and simplicity from all these communities like the change of colours of earth in the cycle of seasons. To justify the fact, there is a proverb in this region like this.

"Luhuk kaate Tadam Lekhaan Enez
Luhuk Kaate R-al Lekhan Serenz."
Or
"Senge susun,Kaajige Duran"

It means the rhythmic dropping of the legs is 'dance' and the sweet verbose is 'song' whether remaining under the strain of cultivation in filthy atmosphere or at the time of collection of the different forest products. In every step of life, these tribes fill the atmosphere with their melodious tunes of songs. They never concentrate particularly on songs with a practising mood but song is their natural goal and destination. In addition to the songs, their dance steps descend with the change of seasons. Of course, on the eve of one ceremonial festivals, all the tribes are boozed in song and dance, but each tribal group has its own songs having separate dance patterns. For example, the Santhal community in the month of Jan-Feb are involved in "Dahar dance and song', the Bhumijas enjoy 'maang Buru Susun'and the Kolhas enjoy 'Mage'. Similarly, while the Santhal on different festive occasions are engrossed in the dance like Baha, Lagne, Rinza, Dasain, Bhuang, Budhigaadi, Dhanged, Singurai etc. The kolhas are engrossed in the dance and song of Hero, Baa, Aandi, J-manami and the Bhumijas are engrossed in Hadi Susun, Aanaadi, Dasain etc, but all the sings and dances are not homogenous having no affinity. All the songs and dances of every tribal community have own individuality and are elementary for the culture of each community.

Dance and music may be created for celebration of institutional and traditional festivals, but it can not be discarded with the fundamental ethos of all castes and creeds that is contained in it. In case of institutional marriage ceremony, for example, the Santhals perform 'The Doun Dance' where as the Bhumijas perform 'the Aanadi', the kolhas 'The Aandi' and the Kudumis 'The Nachani'. Likewise, while observing the 'Gamhan Purnima', the Santhals perform 'the Rinjhaa' and the Bhumijas 'The "Kaaraam'. This is how there are different kinds of dance and music within each community of every casts and creed those are performed throughout the year. Of course, an amusing fact is that there may be prohibition of the entrance of the people of one community in the participation of the group dance, song or festivals of other communities, but it is remarkable to note that in the folklores of all the tribes the forest products, the forest surroundings and the analysis of ecological ideology are well-marked.

The tribals and the indigenous people make their children familiar with Nature in the tradition of the folk literature.

Ecological ideology generally finds its footprints in many lullabies meant for children. Sometimes the mother singes to the child at the bed-time.

"Nindaa Nindaa Taapi Taapi
Aale bay Da jaapid ked
Aalaam Hiju Banilaa Siaal
Aale Boy da jaapid keya
Aalaam Hiju Baru Baanaa
Aale Baabu Da jaapid ked
Nir jamme Baana Siaal ka."

It implies, "My dear son. You sleep. If your cry breaks the silence among the cold atmosphere and serenity of the forest, the tail-less fox of the forest and the bears or tigers of the bush will go away and cordial relationship between man and beast will be at stake. My son is a good chap; he is not naughty like the beasts of the forest".

In this song there is the illustration related to the beasts of forest and within it some special words like *Baanila Siaal, Burubaanaa* are used by the grandpaa and grandma since childhood. They have been taught to love the forest in these tales.

In this context, the education of a child about the philosophy of ecological ideology of the tribes of eastern India starts since his childhood.

Many assume that 'Ayurved' is the medical science of the tribals. But there is a hell-and-heaven difference between the treatment of 'Ayurved' and 'tribal-folk treatment'. Perhaps different tribals unconsciously follow part of the Ayurvedic treatment. But the methodology written in Ayurvedic science is only a meagre part of their treatment method. The methodology of treatment followed by the tribal community may be divided into three categories:

1. BHESAJA CHIKISTA (medicinal treatment)
2. SABDA CHIKISTA (sound treatment)
3. DAYBA CHIKISTA (spiritual treatment).

In short medicine treatment system of these tribes is used for different diseases as per the 'Tithi, Baara, Graha, Nakshyatra' and their medicines are prepared by a mixture of dung, urine, bone, flesh of beasts along with the flora, metals, stones, etc. collected from forest. This system also exists in Ayurvedic treatment.

Secondly, 'Sabda Treatment' is nothing but the treatment through chanting of 'mantras' by using specific words. We all bear a sound within our body since the beginning of the creation. Those are divided into four– (a) Paraa, (b) Pasyanti, (c) Madhyamaa and (d) Baikhari.

Knowing this fact, the wise saints and monks had created certain mantras. Different communities of tribal sect have also adopted similar method.

Thirdly, 'Daibya Treatment Method' of the tribals bears their own speciality. This method is known as 'Air Treatment'. Different communities of the tribals have also adopted this method observing the movement of different kinds of air i.e. 'Praan Air, Aapaan Air, Samaan Air, Udaan Air and Byaan Air' within our body.

All these are the own medical science of the tribals. Different forest products are the basic elements of this treatment. The names of the disease and medicine of this treatment are quite different from that of this Ayurveda. In the tribal scripture, the knowledge about the quality of the medicinal plants are also different from Ayurveda. The chemicals like mercury and phosphorus are not in use in the method of this treatment. There are detailed descriptions about the system of medicinal treatment found in the folksongs of different tribal community. If these are collected and published, it will throw light on the philosophy of ecological ideologies of the tribals. We see, there are different types of fevers and their elaborate descriptions of fevers and the medicines are found in the folklore of the Lodha community.

The folklore tells about the insane fever of the frustrated lover:

Bandhaake Aaila Jwar
Kaahaan Pawba Ojha Ghar
Bandhe Debon Naagaila Sikad
Ge Sajane.

The song means to say 'My darling is suffering from insane fever. The doctor is not available now, I know that the unbeatable medicine of this fever is the root of tree named *Naagaila.*

There are innumerable songs filled with the thought of ecological ideology. The consciousness of Indian ecology always appears in these songs. It was reflected during the British rule in the District Gazetteers and the official reports. These are not being analysed after the independence of India. Without trying to understand the tribal culture, the non-tribal officers from the outer area have attempted to impose the pressure of their own culture upon the tribals. They have also tried to keep them away from the collective memory about their folk songs and folk literature while making the tribals educated and civilized. There is a cultural ragging on the rich civilization full of forest resources which belongs to the tribals forever. As a result of which the tribal people sing :

Jami Lilo, Baadi Lilo
Dhaan Chaal Lilore
Ghar Lilo, Duaar Lilo
Aarai Lilo Maanre
Ban Khandi Paatane De Saanre……

This song implies that the 'non-tribal officers have cleverly pocketed the forest property, house and land of our region. Our cultural heritage has gone to dust. So my dear brothers, oh the lovers of my land and forest, keep ready your bows and arrows- the tribals tell about the cause in these lines.

'Noaaga Odisha Haanaagaa Bihar
Hantegaa Baiha Bengal Disam Kana
Jaanaam Sisam Akaare Han Ge Baana
Janam Disam Aaya Etaa Jaarire Menaa
Etaa Jaarire Kaalmi Lekhha Senaa'.

The song says 'The mother land of the tribals living in the forest has been turned into a maid-servant in the civilization of Bihar and Bengal. The forest lands have been raped today.'

The cultural history of the tribals and the indigenous people is not being written properly, with all information correctly. Because they do not have the opportunity to express their creative talents though they possess it vehemently. Of course, most of them cannot write in regional languages like Hindi, Bengali, Oriya or Assami, but they are able to write in their own language thousands of Jhumar, Daun song, Laagne song and Rinjhaa song. So one can not say that they lack creative potentiality. Moreover, no regional government is stepping forward for the publication of these folklores. So the tribal poet sings.

Desh taa Ta Swadhin Hoilo
Du Kudi Dash Bachhar Gelo
Baabu Bhaayay Sakal Lute Lilo
Take Churain Khailo
Tar Chhadain Lilo
Aar Tuin Gaamchha Parhaa
Take Thepo Dilo
Din Gelain Banta Dhari
Kaday Sarpat Kharai Mari
Dlku Pelaah Jaanum Jhaanti Bhay
Taatka Tai Dhukaabekre Dhalay.

It means 'The tribal people are suppressed in the trap of the officers coming from outside even so many years after independence. They are still cultivating in the sun and the rain and dedicating their lives in the services of those officers by collecting forest products for them'.

To know the life philosophy of these tribal people, one has to enter into the arena of their life style in the forest and their tribal culture in this context. They have expressed their life philosophy and ecological ideologies in various songs to pronounce their education and culture. So it is the call of the time to rewrite urgently their historical background and to analyse their ecological ideologies.

8

PROTECTED FOREST RULES AND HO SOCIAL PROTEST

ASOKA KUMAR SEN

Contextualising the forests as historical and empirical discourse is a growing trend. Forest as a resource whetted some studies, which focused it as a power regime leading to the contest for rights between the state and forested communities.[1] Then there are works, which study forests in the context of social ecology and environment. Some of these highlight the colonial and post-colonial state policies on the forests and the contesting ecological ideologies.[2] Though differing in focus these studies however explicitly or implicitly argue that an extensive investigation is necessary to create a more informed forest-related corpus. With this idea in mind this paper regionalises forests by taking up the specific problem of the Ho tribe of Singhbhum in the context of the colonial forest rules. In doing so the paper first draws the ideological setting, secondly it studies the formulation of the forest rules and it ends up by elaborating the content and specificity of the tribal social protest that these caused.

The Ideological Setting

Though the state management of the forests was systematised since the 1860s with the passing of the Indian Forest Acts, it took three more decades to invoke the same to formulate local rules. That the natural resources of Singhbhum, particularly Kolhan's *sal*-rich forests, had tremendous revenue potential seemed to be triggered by the railways that extended their network in Singhbhum since the 1890s.[3] A local forest official admitted 'In early days no one thought of the jungle one way or the other because there was so much of it that it was a positive nuisance. But now it is a valuable property.'[4] Soon the British was found to explore rich mineral reserves, some of which were located in the forested parts of the district, and also forest produce like timber for construction, tamarind and *sabai* grass.[5] The lure to appropriate revenue no doubt made the British officials protection-prone.

While dabbling with the idea of protection the British learnt about the close link of the tribals with the nature and their historic forest rights. They thought that to protect them would be in line with their paternal rule of the Kolhan tribals. But this altruism was influenced by daunting experience of tribal ire when their rights were transgressed. In this background W. B. Thomson, the Deputy Commissioner of Singhbhum, and later his superior in office recalled the Birsite revolt.[6]

However, the British could not afford to be cowed down by the fear of a rebellion. As the memory receded with time parallel pockets of autonomy in the form of the Village Forests enjoyed by villagers under the control of village and *pir* (cluster of villages) heads appeared to be 'essentially inconsistent with the ultimate ownership of the state.'[7] Aggrandisement was clear also in British attitude to bring private forests of zamindars under state control.[8]

This lure of power was screened by the will to preserve the woodland from the profligacy of the forest people and unscrupulous private contractors ushered in Kolhan by the local zamindars.[9] However, generally speaking fingers were pointed to the tribals. In 1901 the Commissioner of Chhotanagpur criticised them for recklessly denuding the state forest of valuable timber.[10] A forest official believed that 'If left to their own devices the aboriginals will clear everything in less than one generation.'[11]

However, a forest official blew off the cover that hid the skeleton of government's connivance when he observed: 'But I have no knowledge as to how the Calcutta firms managed to get their leases.'[12] While the officials posed that the real purpose of state forestry was 'not revenue but preservation of jungles for the use of villagers,'[13] imperial interest in maintaining the supply of timber was affirmed.[14] Obviously for revenue private companies were given monopoly rights to appropriate *sabai* grass.[15] The DC drew attention to the tremendous material prospects of the forests with the words 'I consider DFO would do well to bear in mind the possibilities of large developments in pit props and shoring timbers, and in fuel and charcoal, consequent upon the great developments in mining work and mining settlements which will arise in the next few years.'[16] A welcome government leased out large tracts of protected forests for the mining purposes.[17] Denudation of the forests was the natural result.

Yet the forest officials averred that the pursuit of state forestry was ecological conservation. Awareness of forest-ecology nexus was growing in India after the conservation debate of 1878. The Chhotanagpur administration, and through them the district administration, was awakened to the disastrous effects of deforestation on the 'climatic conditions and the rainfall of the country'.[18] They had the knowledge of the warnings sounded by Fr. Hoffman of maintaining the needed forest cover to ensure sufficient rainfall to retain the moisture of soil.[19]

Formulation of the Forest Rules

This ideological setting escalated the concept of state forestry, which had begun, with the reservation of the forests in Singhbhum. Outside the reserved forests all forests in the Ho dominated Kolhan were known as the Village Forests, which were enjoyed by the raiyats subject to the control of the village headmen. Under the Notification No. 3375 For. of 5 September, 1892 all unreserved forests were made the Protected Forests by the government. Invoking section 31 of the Forest Act of 1878 the local administration then issued the Protected Forest Rules in 1895. During Craven Settlement of 1895-97, 58 forest blocks were demarcated and in 1903 the remaining areas were declared to be the Village Forests. However, due to the loss of revenue the district administration considered the change to be 'undesirable' and framed a set of rules called Twidell's Rules in 1903 'to regulate the disposal of trees and other forest produce', which was duly approved in 1904. Later 14 more Protected Forest Blocks were carved out of the Village Forests.[20]

At the same time after years of hesitation the government finally prepared the record of tribals' forest rights as required under section 28 of the Act of 1878. When this was done by L. B. Burrows and A. C. Das it was found that the rules contravened the *pir* people's existing rights of collecting timber, forest produce and of grazing cattle subject only to the control of their village (Munda) and *pir* (Manki) heads.[21]

The British subsequently brought other changes. They allowed private companies to collect and sell *sabai* grass and minerals. It meant that the government was pursuing the dual policy of allowing others the right to market forest and mineral resources for profit and earning money from the deal, while the tribals were only accorded limited usufruct. The Notification of 24 March, 1922 imposed grazing taxes of 8 (subsequently reduced to 6) annas for each sheep and goat over the number of four. They argued that sheep and goats caused greatest damage to the forests and so they were the greatest conservation hazards.[22] Second argument was that Kolhan tribals kept cattle much over their requirement.[23] Next, none other than the Deputy Commissioner himself believed that this tax would not be a burden on the community as they were affluent.[24] In this way the forest rules shrinked Manki-Munda's traditional power and jurisdiction and encroached into the customary forest rights of the tribals, thereby creating a situation of conflict and confrontation.

The Protected Forest Rules and Ho Protest

More famous in history as the *Larka* (fighting) Kol, the Kolhan Ho had a tradition of anti-British struggles. During 1819-20 they resisted the efforts of the chiefs of Porahat, Seraikela and Kharsawan to subdue them with the British help. They fought against the British bid to subjugate their territory in 1836-37. They rose against foreign domination during 1857-59.[25] But later they seemed to rationalise with the foreign domination and live peacefully under British rule. But from the colonial records it becomes evident that discontent was simmering within against the transgression of their customary rights. We learn of tribal protest, seeking to ameliorate their grievances within the framework of British rule and not taking recourse to the armed way as they did before.

This Ho protest followed two modalities. First was that of disobedience and wilful breaking of forest rules. Such cases of law-breaking respectively for 1914-15 and 1915-16 were: injury by fire 30 and 30; unauthorised felling or removal of forest produce 77 and 49; unauthorised grazing 5 and 1; while other offences were 6 and 26. While the total of 106 for 1915-16 showed a decline from 118 of 1914-15, but these registered a marked rise from the average of 88 for five years ending 1914-15.[26] Moreover, the village heads of Bila, Sonua and Regarbera ignored the official forest boundary and included portions of the reserved forests within their villages.[27]

Breaking law was however hazardous. In Singhbhum the Ho had for almost a century acclimatised themselves to the legal way of functioning by moving the authorities for the mitigation of their grievances. Naturally therefore the Ho villagers submitted petitions to the officials. In 1906, 33 Mankis, Mundas and villagers sent a petition to the Commissioner against the formation of some more protected forest blocks. They objected to the official decision of taking out the village forest areas from the supervision of the village-pir heads and placing them under the management of a Special Officer. They were alarmed also at the adjoining and settled lands of the villagers being resumed and included within the protected forests and a strict supervision being imposed on them.[28] They prayed that the Protected Forests be kept and managed by the Mankis and Mundas, *gora* lands of tenants be not resumed and included within these and no change be made in the established system of Protected Forests.[29]

Interestingly the petitioners refuted the patent official charge of tribe as the destroyer of forests. They considered the forests as nature's gift to them and as such their being invested with the right to appropriate the resources, contesting thereby British attempt to usurp the same.[30] However, this contest for power on the strength of their customary rights seemed to be more theoretical than real when they argued that if settled *gora* lands were resumed 'they will be deprived of the earning by which to pay the rent of their own lands.'[31] The willingness to pay rent was an act of civil obedience.

More deep rooted was however the projection of their identity of a cultivating people. They contended that due to deprivation of grazing spaces the growth of their cattle would be stunted and their cultivation would suffer; they would be denied trees from forests which they used for making ploughs; if the *gora* lands were resumed they would have to sell paddy for paying rent, depleting their reserves which would ultimately affect their cultivation and lead them to starvation.[32]

The arguments were woven around an existentialist logic that highlighted the predicament of a marginalised cultivating people. Marginality was more pathetically projected when it was said 'fire is the only means of their protection from cold. If they do not get easy access into and free supply from the forest they will have to die of cold.'[33] This notion of self disputed the generic official notion of plenty and affluence expressed by the DC that 'the villagers pay very low rent and cesses, and are exceedingly well off.'[34]

However, they insisted on a peasant-culture deeply rooted in the forests. Petitioners observed 'That the inhabitants of Kolhan are naturally fond of jungles than populous towns; their habits and customs are suited to living in the Jungle; and if the Protected forests which are under their control be taken away from their hands they will be out of their elements and will feel the greatest hardship.'[35]

Basic to this cultural difference was also the varying aesthetics of life which followed what a recent study characterises as 'the Culture of Present'.[36] Contending the glib official logic of preserving the jungles for 'the future generation', villagers insisted 'The present want must be first supplied and then provision for future want should be made.'[37] This was a common man's prudence developed from the logic of survival to contend the colonial castigation of the tribe as profligate.

The Ho community was more widely agitated when the District administration imposed grazing taxes. Against this the villagers of Kolhan filed a petition to the DC which carried the signature and thumb impression of more than one thousand raiyats, being presented by prominent Mankis and other social leaders to the district head himself.[38] About the highly charged atmosphere and official alarm Scott remarked 'I have recently had to deal with the widespread complaints from the tenants of the Kolhan government... very large number appeared at the kutchery in Chaibasa to lay their grievances before me. While the attitude of the raiyats has so far been perfectly law-abiding the matter is clearly one upon which they feel very strongly, and unless prompt steps are taken, might easily be magnified into a cause of trouble through out Kolhan.[39] A delegation of ten Ho social leaders met Sheikh Abdul Hakim, the second officer, on 2 February, 1924 and represented their case obviously recounting their arguments of the petition.

After briefly reminding the government that the forests of the entire Kolhan Government Estate were managed by the Mankis and Mundas formerly, they charged that government had deviated from the avowed policy of protecting the forests for the welfare of the people because they were not allowed 'to take wood and the jungle produce free of royalty from those protected forest blocks although we were promised to get wood and other jungle produce for our use free of royalty'.[40] Second argument was that the taxes were historically unsupportable, as they had 'never paid grazing tax to Government'.[41] After thus arguing within the ruler-ruled network of relations the petitioners claimed usufruct as a matter of right for the free services rendered by them in the form of extinguishing forest fire and clearing the lines.[42]

Petitioners however put sharper focus on their specific peasant-tribe identity once again. Assertion of peasant identity was made through the evocation 'We have no other means except cultivation.' This was reinforced by such statements that payment of grazing taxes would hamper their cultivation generally and particularly of those who had 'large quantity of land' requiring 'large number of cattle for cultivation'; they had to maintain 'cows and cow buffaloes, sheep and she goats for breeding purposes' without which cultivation would suffer at a time when market prices of cattle were very high; they used cow dung for manuring which required a large number of cattle.[43] They were thus particularly contending the official blame that the Hos were maintaining excess cattle. Moreover, they suggested that the above Ho agrarian practices had been evolved not to promote agriculture alone but to perpetuate the social mechanism of homogenisation. They were pointing to the practice of lending bullocks, cows and buffaloes and cow-dung to those who had less or none by those who had them in plenty. In return they paid goats and sheep for feast to the village community.[44]

More innovative was the articulation of the positive and negative aspects of Ho social identity. It underlined their custom according to which 'during our marriages we have to give numbers of cattle towards *Pan* (italics mine. *Pan* means bride price). If we do not keep sufficient number of cattle for this purpose, it is impossible to meet the demand of *Pan* during the marriage of our children'.[45] Then again 'we during our illness use to make Pujas by sacrificing our sheep and goats instead of using medicine...The goats and sheep are also used for performing marriage and death ceremonies and in the festivals.....We do not keep cattle for selling Ghee or Milk.'[46] This way they were insisting their difference from those who used medicine and sell ghee or milk. This specificity of Ho self-image was thus invoked as a tool to safeguard their forest rights.

Conclusion

The DC confabulated with the delegation, accompanied by around two thousand irate tribal villagers, and finally declared the withdrawal of the grazing rules.[47] The impressive and unprecedented tribal protest against forest rules unmistakably showed their concern for the protection of customary forest rights. But what impresses a researcher studying Ho most is both the content of their petition and the logic behind their arguments. Significantly they were arguing on the basis of their communal existence that at once was a defence of historically enjoyed customary rights as well as the projection of their distinctive cultural identity within the Indian mainstream.

Archival sources used in this essay are from the District Record Room, Chaibasa.

NOTES AND REFERENCES

1. D. Hardiman, 'Power in the Forests: The Dangs 1820-1940' *in Subaltern Studies (SS)*, Vol. VIII, Oxford University Press (OUP), New Delhi, 1999; D.Arnold, 'Rebellious Hillmen: The Gudem-Rampa Risings 1839-1924', SS. I, OUP, Delhi, 1986; T. Sarkar, 'Jitu Santal's Movement in Malda in 1924-1932: A Study in Tribal Protest', SS IV, Delhi, 1990; S.Dasgupta, Adivasi Politics in Midnapur, C. 1760 – 1924, SS IV, OUP, (Delhi, 1990.
2. D. Arnold & R. Guha (ed.), *Nature, Culture, Imperialism*, OUP, New Delhi, 1999; M. Gadgil & R. Guha, *This Fissured Land*, OUP, New Delhi 2002; R. Guha, *The Unquiet Woods*, Delhi, 1989: M.Rangarajan, *Fencing the Forest*, OUP, Delhi, 1999; K. Sivaramakrishnan *Modern Forests: Statemaking and Environmental Changes in Colonial Eastern India*, OUP, Delhi, 1999; A. Skaria, *Hybrid Histories: Frontiers and Wildness in Western India*, OUP Delhi, 1999.
3. Letter from C. N. De, PA to the Commissioner to the DC, Singhbhum, No. 804 LR of 1 July, 1907 vide Protection of Private Forests, Fly Leaf (FL), Deputy Commissioner, Singhbhum, Office (DC's Office), Revenue Department (RD), Collection No. (CN) II, File No. (FN) 16 of 1908-9, File Serial (FS) 3, Appendix I, para. 6 & FS 7, *Ibid.*
4. Divisional Conservator of Forests' (DCF) comments in the margin to the Note by H. D. Carey, Deputy Commissioner of Singhbhum, on the Preservation of Private Forests in the Chhotanagpur Division, Singhbhum *vide* Protection of Private Forest, op. cit., para. 4.
5. J. A. Craven, *Final Report on the Settlement of the Kolhan Government Estate 1897*, Calcutta, 1898, p. 15.
6. W.B.Thomson to Conservator of Forest dated 23 April, 1900, *vide* Leasing out the Sabai grass, FL, DC's Office, RD, CN II Forests, FN 7 of 1900-01, FS 2, para. 4; A. Forbes, Commissioner, Chhotanagpur Division to Conservator of Forests, dated 15 July, 1901, Leasing out the Sabai Grass, *ibid.*, FS 3, para. 5.
7. Report of L. B. Burrows, Appendix to DC, Singhbhum's letter to the Commissioner, No.181R of 15 May, 1912, *vide* Constitution of the Kolhan Reserved Forest Division, FL, DC's Office, RD, CN II Forests, FN 1 of 1915, FS 52, p. 19.
8. H. D. Carey's Note and DCF's comments in the margin, paras. 3, 8, 10.
9. Carey's Note, para. 6.
10. Report of L. B. Burrows, para.2.
11. Report of L. B. Burrows, para. 6. See also, Formation of the Kolhan Protected Forest Block, FS 9, para. 2.
12. DCF's Comments, Carey's Note, para. 6.
13. Carey's Note, para. 8.
14. *Ibid.*, para.1.
15. The Conservator of Forests of Bengal apprised the Commissioner of Chhotanagpur 'There is no doubt that from a financial point of view it would be to the interest of the Government to give the forests out on lease.' Lease of Sabai Grass in the Protected Forest. FS 1, para. 4. In fact the sabai lease in the RF to G. Henderson & Co. earned revenue of

Rs. 1,26,000 for three years. W. B. Thomson to Commissioner, 6/10 June, 1901, FS 2, para. 1.

16. Letter of J. E. Scott to DFO, N. 3659R of 20 July, 1923, Rules for the Pasturage of Cattle in government Protected Forests (Revision of the Protected Forest Rules), FL, DC's Office, GD, RB, CN II, FN 16 of 1923 & 5 of 1924, FS 13, para. 26. See also Letter from V. G. Pigott to H. F. Samman, DC, 22 April, 1904, *vide* Application of Mr. V. G. Pigott for a prospecting lease for lime stone in the Kolhan, FL, DC's Office, RD, CNXV Mining, FN 7/5 of 1904 –5, FS 13, para.10.
17. As an example one may see Land Revenue Administration Report for 1928-29, FL, DC's Office, GD, RB, CN XI, FN 19 of 1929, pp.15-20.
18. Cited in Report of L. B. Burrows, para. 2.
19. C. N. De, PA to the Commissioner to the DC, Singhbhum, No.804 LR, Appendix, FS3, para.7, *vide* Protection of Private Forest.
20. Letter of DC, Singhbhum to the Commissioner Chhotanagpur Division, No.1222R of 2 November, 1913, Constitution of Kolhan Protected Forest Division, FS. 61, paras. 2-5.
21. Report of L. B. Burrows, pp. 12-22.
22. *Ibid.*, p. 20.
23. Forest Administration Report 1915-16, Fly Leaf of English Correspondence, DC's Office, RD, CN II Forests, FN 9 of 1916, para. 44.
24. Letter of J. E. Scott to DFO, para. 2 (b).
25. M. Sahu, *The Kolhan under the British Rule*, Calcutta, 1985, Chap. III.
26. Forest Administration Report 1915-16, para. 27.
27. A.K. Sen, 'Singhbhum 1927-30 and the Tour Diary of Ratneshvar Roy', section on trends of Deforestation and conflict over forest right, *History* (A journal of Burdwan University), Vol.VI, No.I, 2003..
28. Petition of the Mankis, Mundas and Tenants of government Kolhan estate in Singhbhum to the Commissioner of Singhbhm, Appendix to Letter of E. A. Gait, op. cit., para. 1-2.
29. *Ibid.*, para. 11.
30. *Ibid.*, para. 9.
31. *Ibid.*, para. 4.
32. *Ibid.*, paras. 4-5, 7, 11.
33. *Ibid.*, para. 8.
34. Letter of J. E. Scott to DFO, 20 July, 1923, FS 13, para. 2 (b).
35. Petition, para. 6.
36. P. Sen, 'Conceptualising Time in a Changing Agrarian Tribal Society,' *Journal of The Asiatic Society of Bengal*, (Communicated)
37. Petition, para. 10.
38. Scott to Hannah, 31 January, 1924.
39. Scott to Commissioner, para.1.
40. Note by Sk.Abdul Hakim, 2 February, 1924, para.1, Appended to DC's letter to DFOs No. 909-11 R of 8 February, 1924, *vide* Rules for Pasturage of Cattle, op. cit., FS 25A, FN 5 of 1924.
41. *Ibid.*, para. 9.
42. *Ibid.*, para. 7.
43. *Ibid.*, paras. 2, 3, 5, 8.

44. *Ibid.*, paras. 5-6.
45. *Ibid.*, para. 4.
46. *Ibid.*, para. 6.
47. DC's letter to DFOs No. 909-11 R of 8 February, 1924, op.cit.

9

JUNGLE KATAI ANDOLAN (1978) AND SINGHBHUM

CHITTARANJAN KUMAR PATY

Introduction

The Singhbhum area is famous for thick forest. Dalma, Porahat and Saranda jungles are very much noted for not only their natural splendour but they are also famous for their rich timber too. Saranda jungles which are known as "Saranda of the seven hundred hills" covering an area 2,11,840 acres are one of the largest jungles of Asia as a whole and have always been a centre of attraction for nature lovers. It is said that except for few trees, almost all sorts of trees of the world are found in the jungle. Research scholars of Forestry Science have a great desire to visit[1] this Saranda jungles for learning and for improving their knowledge of the forested parts of the area.

Mostly the jungles of Singhbhum are situated within the Kolhan region. There are also jungles in Porahat area which are known as Porahat Forest Division. Similarly Dhalbhum area where iš situated the famous elephant resort of Dalma is under the Dhalbhum Forest Division. All these forest areas are populated mostly by the tribals. The *Ho* tribe settle in Kolhan area while *Munda* tribes including *Hos* also reside in Porahat area[2]. These tribals have been living in the remote jungles area of this District for thousands of years after they migrated from Chhotanagpur plateau in about 10th century A.D.[3]

The tribals who migrated from Ranchi District to the areas of Singhbhum District cleared the jungles and settled here. After coming here, they cleared the jungles and prepared the lands for agriculture. The man who first occupied some lands by clearing the jungles the owner of that land and this first settler is known as *Khunkattidar* which means the original male member who personally cleared the land with his own hands. They had always claimed customary ownership of the land and land based resources within the traditional village boundaries[4]. All these could not be sold or transferred from them. On the other hand, they could not be removed from that particular area also[5].

The tribals who live in or around the village depend for their basic needs on the forest environment. Besides agriculture they eat the seasonal fruits like *Mango, Kathal, Imli, Jamun* etc. The jungle serves the materials like bamboo, leaf etc. for construction of their houses and the wood for their fuel[6]. Moreover they procure honey (hurumusu) from the flowers of some tree and timbers, mushroom etc. Forest also provides the fodder, water and shelter to their domesticated animals. Similarly it provides herbal medicines of the village folk[7].

In almost every village there is a small patch of forest called *Jahara* which is a sacred place of worship for them. Thus it is quite right for them to worship forests because their entire living depend on forest. If forest are destroyed they will lose their source of food, medicine, fruits, fodders for their animals and they will lose their only source of cooking and warming themselves. Due to all these the tribals react whenever there is destruction of forest. They have always looked on forest contractors as a sort of pest who eat into their granary i.e. forest[8]. Since their settlement in these areas the tribals have a right to use the forest materials within their villages. They have khunkatti rights over the jungles lands belong to their jurisdiction and even the Rajas and Zamindars of Singhbhum also had recognised the customary rights of the tribal subjects[9]. It is recognised by law *vide* section 8 of the Chhotanagpur Tenancy Act[10].The *Khuntkattidars* of Ranchi and Kolhan area of Singhbhum District enjoy the right to reclaim forest within the villages without any permission of the Government[11].

Jungle Katai Andolan

Now the question arises how and why the tribals revolted and started to fell trees of the jungles when they loved these jungles and these jungles are their livelihood and their very existence. There are many reasons behind it, if we sincerely study the whole affairs of forests and the tribals since beginning of the colonial rule to the independence:–

There are different reasons behind the forest felling movement. The most important reason was the government's forest policies of 1894 and 1952 which debarred them from their customary rights to the jungles. The tribals regarded themselves as the lord of the forest were turned into a subject through officials process and were placed under the forest department. The traditional rights of the tribals were no longer recognised as rights. In 1894 they became "rights and privileges" and in 1952 they became "rights and concessions". Now they are being regarded as "concessions"[12].

Due to promulgation of all these laws, the adivasis who lived in the forest were forced out of their *Khuntkati* villages. They left behind them the burial place (*sasan*) of their ancestors with huge stone placed on the graves (*sasandiri*), their worship places (*desaulis and sarnas*), the paddy fields which they had made and the tamarind and mango trees which they had planted as signs that once they lived there. Thus the eviction from the jungles and infringement of their customary rights over the jungles were resented by them: forced them to rise against the government[13].

During rainy season from mid July to mid September, when the chances of employment opportunities are very limited, the tribal fell trees and sell these for their livelihood. At the same time timber contractors also incite them to fell trees for their personal profit. This is one of the causes of felling trees movement[14]. On the other hand the closure of *Chiriburu* Quartzite mines near Chakradharpur, *Roro* Asbestos mines near Chaibasa, pelletising plant at Noamundi and the closure of NMTC which purchase the Iron ore made thousands of people unemployed. This unemployment increased pauperisation and forced them to look for alternate means for survival. Some of them migrated to the brick kilns in search of jobs out side state while others took recourse to illegal felling of forest for the purpose of cultivation or for selling of firewood or timbers[15].

Due to reservation of forests, the tribals have been constantly harassed and exploited by the officials of the forest department. After the creation of reserved forest area the tribals were removed from their khuntkatti villages within the reserve area. However not all the villagers were removed but some of them were taken over from the civil department and the residents of these villages were kept as bonded labourers by the forest department. These are the so-called forest villages of which the sole malik or owner is the forest department. In this case these tribal are allowed to settle down and cultivate land within the forest demarcated areas and, in lieu of it, they supply every thing right from milk to rice to the forest officials. Not only that, they are forced to work for the department during the forest cutting seasons. Thus the adivasis who lived in these forest villages became virtually bonded labourers to the forest department [16].

Sometimes these officials sexually exploited these tribal women also. This is evident from the fact that in the early 1980s forest Dak Bungalows which were popularly perceived as the vice-dens were targets of arson in Singhbhum when the modesty of tribal women was outraged[17]. Similarly due to the heavy demand of timbers to the Ranchi town which grew fast as an industrial town the timber trade became one of the profitable business of timber mafia. The weekly hats at Hatinghore, Tapkara, Rania, Murhu of Ranchi districts became the centres of illegal timber. About hundreds of saw mills are operating in Murhu and Ranchi, adjacent to Porahat Forest Division. They are the biggest buyers of illegal timber. The official machinery, in many cases, is even afraid to take any action against these organised strength of timber mafia. The villages like Jeraikela, Manoharpur and Goilkera located near the Bombay Howrah main line in Kolhan and the Saranda Forest Division have become the centre of illegal commercial felling at the behest of the timber contractors[18].

The illicit felling tree out side the Kolhan and Porahat areas is largely for the commercial purposes. The pressures of the needs of timber at Jamshedpur and neighbouring districts of Bengal and Orissa also led to felling trees by a well organised gang of criminals who indulged in timber theft. Sometime increased demand for the consumption of alcohol near Jamshedpur, Adityapur and other industrial towns also encouraged felling the trees of jungles. The process of distillation of *Mahua* requires a lot of fire wood. Due to this demand villagers near the jungles fell trees[19].

In the year 1975 Bihar State Forest Development Corporation Ltd. was set up for the purpose of commercial profit from the forests. In this programme they replaced *sal*, *mahua*, *kusum* and other useful and fruit-bearing trees useful for tribals with teak which was more profitable for the department which included khuntkati forest in which tribals have complete rights because their were their sasan and other sacred places which disturbed their religious life also[20]. Thus the tribal communities who depended for their survival on their own trees were replaced by commercial trees as teak. This was the reason that the felling of teak trees during the Andolan by the adivasis was, therefore, a protest against this anti-people programme of the Forest Department. Later on it is reported that the Forest Development Corporation was wrong on this issue[21].

Politically the articulation of economic deprivation took a variety of forms, ranging from the demand for Kolhanistan to Jharkhand to move local and immediate problem like the hunger for cultivable land. Since 1978 onward, there started a strong movement of the adivasis for the restoration of their ancestral land from the Forest Department. They claim that their ancestral "*Sasandiris*" (Stone memorals to the dead) are placed under their jungles. It means their ancestors were buried there. Hence it is the property of that clan of the village. This also led to the anti-forest movement[22].

Besides all these factors mentioned which are directly connected with the forests, there are some other factors also which contributed to the out-break of the tree felling movement. These are large scale legal and illegal alienations of tribal land displacement of tribals due to various projects such as Subarnarekha and Icha Dam project without proper rehabilitation and adequate compensation, increase of unemployment and the frequent drought condition existing in the area. Due to extreme poverty and hunger, the masses who lived in the forest area started to cut the trees of the jungles for selling these for their livelihood[23].

In this way, all these causes led to a militant movement which is known as 'Jungle Katai Andolan' in Singhbhum. This Andolan suddenly broke out in the year 1978. The immediate cause was certain policies of government which were detrimental to the interest of the adivasis. In the year 1973 the Government nationalised the kendu leaves trades; in 1976 it took over the sall seeds trade and in 1978 trade in all the Minor Forest Produce (MFP) was taken over by the Government. In 1975 the Forest Development Corporation (FDC) was formed and a total area of 1.92 lakh hectares were leased out to the Corporation for clearing the forested regions and planting these areas with fast growing and high yielding varieties like teak[24]. From 1977-78 onwards, the Corporation started felling of valuable mixed forest in the Goilkera and Sonua blocks and began to plant the teak trees in those areas. Teak was of no use for the villagers[25].

Provoked by these actions by the Government, entire Chhotanagpur was rocked by violent uprisings demanding *inter alia* the denationalisation of MFP and the closure of the work of FDC[26]. The agitationist first started obstructing the operations of FDC. They gheraoed (surrounded) and assaulted the forest officials. On 17th June, 1978 about 60 Adivasis led by *Mora Munda* of *Darong* village hijacked a jeep which was carrying some forest official. He kept these officials in confinement for several hours at *Siyankal* village. These Adivasis demanded that the officers had to give in writing that the work of the Corporation would be stopped. Similarly on 25 June, 1978, the agitationists beat up some forest officials for not closing down the corporation[27]. All these angered the Government and the result was Simdega Firing. On 4th August, 1978 when massive crowd of Adivasis protested at Simdega against State take-over of MFP trade, the police fired on the innocent Adivasis which resulted in the killing of one person. It is to be noted that it was during the time when Jharkhand movment was also at its peak. This incident flared up the Adivasis as well as the leaders of the different Jharkhand parties. The result was that on 17 August, there started a large demonstration. 2000 Adivasis with traditional bows and arrows demonstrated in front of D.C. office at Chaibasa in which Deven Majhi and Shailendra Mahato of 'Singhbhum Mazdoor Union' submitted a petition to the D.C. demanding other things as well as immediate closure of FDC and release of Mora Munda who was arrested for detaining the forest officials[28].

All these firings and ejection of the tribals from their home lands, these jungles, forced the Adivasis to start movement. Thus the jungle Katai Andolan for reclaiming land for cultivation was first started by Machua Gagrai in the Karaikela area who had organised about 67 villages for this movement.

These agitationists now started reclaiming the lands which once belonged to their ancestors but had been taken over by the Forest Department. They pointed out that *Sasandiris* (burial stones) of their ancestors found in forest areas were clear evidence that those lands once belonged to them. Sule Purty of *Seomari* area of *Songra* range of Porahat also organised the Adivasis of the area. It is significant that this is the area where Birsa Munda had taken shelter before his arrest by the British. Latter on, *Magi Dorai* and *Jorgo Munda* in Serengda area also joined this movement.

In the next phase there emerged a new leader who was Sailendra Mahato. His participation in the movement changed the direction of the movement also. In this phase, resistance to teak plantation which forest officials had planted became the symbol of the movement. The Adivasis changed their strategy of the Andolan and this Andolan became known as the 'Saguan (teak) Katai Andolan'. On 2 October, 1978 after a public meeting a crowd of Adivasis ransacked the teak nursery of *Serengda* and destroyed several huts, a pump set and all the teak saplings. On 3rd and 4th October these agitationists destroyed teak nurseries at Buru, Goilkera and Banskatta also. During the time about eleven teak nurseries in Sonua and Goilkera areas were destroyed by the adivasis which caused the loss of about 7.6 lakh rupees to the Corporation. In this way, the Jungle Katai Movement spread like wild fire in the forest areas of Sonua, Goilkera, Manoharpur, Bandgaon and Chakradharpur Police Stations. All the Teak plants were uprooted, huts in the nurseries were burnt down and costly pump sets belonging to FDC were destroyed. Not only that, after destroying the Teak plants the adivasis began to start ploughing the area for cultivation. The result was that all the roads in these forest areas were closed down and operations of the Forest Development Corporation came to a standstill. Again on 9th October, 1978 about 7000 villagers assembled at Goilkera market with their traditional weapons and a meeting was held in which Deven Manjhi and Sailendra Mahato and others addressed the people. They sent petition to the Prime Minister and Home Minister stating the reasons for unrest in the forest area and requested them to stop the Teak plantation in the Porahat Forest Division. Similar meetings were held at Sonua, Longo Kuira etc. And thus the agitation spread all over Singhbhum[29].

Due to all these movement government took action and thus many leaders were arrested. Some leaders like Sailendra Mahato went underground. Sec.144 of C.R.P.C. was imposed throughout the forest areas on 2 November, 1978. In spite of all these still the Adivasis protested against the planting of Teak in the jungles. Again, they gathered for a meeting at Ichahatu village near Goilkera block but the police without giving any warning indiscriminately fired at the crowd, killing Maheshwar Jamuda of Nugri village and wounded several people. Similarly, again, the police fired 12 rounds at Serengda weekly market on 25 November, as a result of which three people died on the spot and many were seriously injured. The police was searching for the main leaders of the Movement and on 1st December, 1978 Devendra Majhi, the leader of Forest Andolan, was arrested at Chakradharpur. They were also searching for Sailendra Mahato at Lonjo village but he was not found there. Having not found Sailendra Mahato the police harassed the villagers including women also. When the Munda (village head) of the area complained the matter to the higher authorities he and some others were arrested. There were wide condemnations of these firings at the national level and repression by the police in Singhbhum. A group of representation led by Sukhdeo Banra-President, Kolhan Kisan Samity, met the Prime Minister on 22 December, 1978 and submitted a memorandum signed by seven M.Ps of the area for demanding action against the officers responsible for firings[30].

In spite of all these repressions the Jungle Katai Movement continued with vigour. On 25 May, 1979, a group of Adivasis women gheraoed the forest ranger at Toklo village. In the meantime on 26 May police succeeded in arresting Sailendra Mahato. This agitated adivasis. In the meantime an adivasi villager who was taking some wood for making plough was fired at and arrested. This led to another agitation. The adivasis of Jonko village burnt down the forest bungalow of Bandgaon in retaliation of the arrest of the poor adivasis. In the meantime in April 1979 Horo Group of Jharkhand Party dissociated themselves from the Forest Andolan under the leadership of Jharkhand Mukti Morcha. The sole leader Devendra Manjhi of JMM led the Jungle Katai Andolan which took a new turn.

As the JMM had taken over the leadership, the Jungle Katai Movement gathered momentum and started spreading to wider areas under the banner of JMM. Again there started public meeting and black flag demonstration at Chakradharpur on 15 August, 1980. In the meantime on 4 September, 1980 about 158 tribals were arrested at Goilkera on the charge of cutting the forest. These tribals claimed that they were clearing the land belonging to their ancestors and not that of the forest department as there was many sasans found in that area. Thus whenever jungle katai movement gathered momentum, the repression was started by the police. This was the reason why Bihar Military Police (BMP) was deployed throughout the forest area but these forces harassed the adivasis in several ways. Due to the repression of police a meeting was organised at Gua on 8th September, 1980 in which police fired at the innocent adivasis which resulted in many casualties. After Gua firing a reign of terror prevailed. Mora Deogam and Machua Gagrai were arrested on 19 September night. Many people fled out from their villages and took shelter in deep forest[31].

The Jungle Katai Movement became a part of Jharkhand Movement and Kharkai Bandh Sangharsha Samiti. Lastly, Gangaram Kalundia who was the leader of Kharkai Bandh Sangharsha Samiti was brutally killed by the police. This also enraged the adivasis. They held many meetings all over the Singhbhum district which condemned this killing. In the meantime JMM was divided into two halves led by Sibu Soren and Devendra Manjhi. This also weakened the movement. On the other hand, due to chaotic condition of the Movement, there started a large number of looting of forest by felling the Saguan trees by the contractors and merchants with the collusion of the forest officials. Hundreds of trucks of valuable timber were daily smuggled from the jungles to Chaibasa-Chakradharpur and other places. The poor adivasis were used by these looters and thus innocent tribals were blamed for this and it defamed the Jungle Katai Andolan[32].

However this movement later on gradually became weakened because the leaders did not care for the victims of police atrocities. This discouraged the people. At the same time some leaders left this movement and joined some other party for election purposes. Moreover due to continued harassment by the police and court cases many people withdrew themselves from this Movement. On the other hand some of them formed another party named Jharkhand Kranti Dal which started a movement against eucalyptus plantation in Dhalbhum area. This also diverted and weakened the Jungle Katai Andolan. Lastly, due to lack of the proper leader, like the Sardari Larai this Andolan also came to an end.

It is undeniable that though the Movement came to an end it was a struggle of the adivasis for their self-identity in which not only male members but women and students also participated. A question still remains that if the government continues to force the tribals out from their original habitat through forest laws and plantation of teak trees for commercial use, what will be the fate of the poor tribals ?

I am grateful to the Mathew Memorial Library, TRTC, Lupungutu, Chaibasa for the collection of information.

NOTES & REFERENCES

1. P. C. Roychoudhary, Singhbhum District Gazetteer, Patna, 1958, pp.107, 446.
2. Mathew Areeparampil, *Jungle Aur Adibasi: Soshan Ke Shikar* (in Hindi) Tribal Research Training Centre, Chaibasa, 1988, p. 2.
3. A. D. Tuckey, Final Report on the Resettlement of the Kolhan Government Estate in the District of Singhbhum, 1913-1918, Patna, 1920, p.17.

4. L. K Mahapatra, Customary Rights in land and Forest and the State, in Mrinal Miri (ed) *Continuity and Change in Tribal Society*, Shimla,1993, p. 90.
5. Areeparampil, op.cit., p.7.
6. Walter Fernandese, *Forest, Environment and People*, Indian Social Institue, New Delhi, 1983, p.13.
7. L. P. Vidyarathi, Role of Forest in Tribal Life in S.P.Sinha (ed) *Tribal And Forest*, Government of Bihar-Welfare Dept., Bihar Tribal Welfare Research Institute, Ranchi, 1987, pp. 323-328.
8. S. Muhammad, Forest and Tribals in Bihar in S. P. Sinha (ed) op.cit., pp. 57-58.
9. L. K. Mahapatra, Customary Right in Land and Forest and the State, in Miri (ed), op.cit., pp. 89-90 .
10. C. B. Prasad, Final Report on Survey and Settlement Operation in the District of Singhbhum 1958-1965, The Supdt. Secretariat Press, Patna, 1970, p. 111.
11. M. C. Sarkar, Customary Right in Land and Forest of the Tribals in Chhotanagpur-Santhal Pargana Region of Bihar in Miri (ed), op.cit., p.101.
12. U. N. Dhebar, Report of the Scheduled Areas and Scheduled Tribes Commission, Vol. I, Delhi 1961, pp. 125-130.
13. T. S. Macpherson, Report on the Operation for the Preparation of Record of rights in Pargana Porahat, 1905-1907, Calcutta,1908, op.cit., pp.160-161.
14. Amarjeet Sinha, From Popular Protest to People's Participation Towards Revolution of a Participative People's Programme for the Tribals and Forests of Singhbhum District of Bihar, 1988, pp. 4-6 (unpublished).
15. Areeparampil, Forest Andolan in Singhbhum, in S. Narayan (ed) Jharkhand Movement ASAKA Publication, New Delhi, p. 70.
16. *Ibid.*, p. 67.
17. Sarkar, op.cit., p. 103.
18. Sinha, op.cit., p. 8.
19. *Ibid.*, 9.
20. Vanki Larai, Peoples Union for Democratic Right, New Delhi April 1982, p. 20, Areeparampil, p.71.
21. Areeparampil, op.cit., pp. 68-69.
22. Sinha, op.cit., pp. 7-8.
23. Areeparampil op.cit., p.70.
24. Areeparampil, op.cit., p.71.
25. Agapit Tirkay, Jharkhand Movement, Others Media Communication, New Delhi, 2002, p.108.
26. Areeparampil, op.cit., p.71.
27. Tirkey, op.cit., pp. 108-9.
28. Areeparampil, pp. 71,74.
29. *Ibid.*, pp. 74-76.
30. *Ibid.*, pp. 76-77.
31. *Ibid.*, pp.78-81.
32. *Ibid.*, pp.81-89.
33. *Ibid.*, pp.93-94.

10

JARAWA

The Negrito Forest Dwellers of the Andman Islands

CHUMKI PIPLAI

Submission

In the Andaman and Nicobar Islands 84.01 per cent (i.e. 6,930 square kilometre) of land is covered with forest (Penguin Year Book-2005). The territory demarcated as Jarawa Reserve is 638 square kilometre. This area has been retained for exclusive use of the Jarawa tribe considering their mode of subsistence, way of life, social practice and cultural values. But this territory is often used by the poachers and intruders. The outsiders are encroaching upon their land, enjoying forest products and fishing in coastal water. This situation has imperilled their life supporting system as well as survival.

Present paper strives to deal with the life style of the Jarawa people in their forest environment.

The Jarawas are the Negrito forest dwellers. They were called by the Great Andamanese as 'Jurua' or 'Jarawa' means 'other people', and called by themselves as 'Aong' means 'man'.

In the long past, during the first penal settlement in the Andaman Islands by the British Colonizers (1789-96), The Jarawas behaved in a friendly manner with them. But after establishment of second penal settlement during 1858, Great Andamanese became more close to the British colonizers. The Jarawas were the traditional enemies of the Great Andamanese, particularly of the Aka-Beas. So, their apathy grew towards the British Settlers. According to Portman's statement (1899) -- "On our arrival the Jarawas were quiet and in offensive towards us, nor did they ever disturb us until we took to continuous molesting them by inciting the coastal Andamanese against them. After the few years of this disturbance, the life of the Jarawas became very hard and in retaliation they began to attack us''.

In 1949, due to the refugee rehabilitation near the Jarawa Settlement, a conflict started between the Jarawas and non-Jarawas on using the forest for survival, which was traditionally used by the Jarawa people. In 1957 the area occupied by the Jarawas was protected as Jarawa Reserve. But after that, different developmental programmes like settlement of a large number of people from outside the island, forestry operation, building roads etc. disturbed the Jarawas. Hence, incidence of violent conflict between the Jarawas and forest staffs persisted. In 1974, some personnel of the Bush Police Force established a friendly relation with a section of the Jarawas. Since then the groups of Government officials including policemen, doctors, administrators, anthropologists used to visit them. The relationship took a new turn after the admission of a Jarawa boy in the Port Blair Hospital. The message of friendship was reached to the Jarawa people through the boy. After that, the Jarawas visited the settlement of the non-Jarawas and were noticed to accept other foods, garments and several items offered to them by the non-Jarawas. In consequence of that some behavioural changes had been observed among them. Some people felt that the Jarawas were suffering from shortage of food and different ailments.

During 2001 and 2002 a scientific investigation was undertaken by a composite team of different disciplines to find out the causes of change in the behaviour of the Jarawas.

Present author was one of the investigators within the study team. Investigation was carried out in two territorial region. Tanmad area (at 5 locations) in the summer time and in the Thidang area (at 2 locations) during monsoon period, in the year 2002. The following salient information presented here had been collected during these investigation-periods.

Salient Information

The Jarawas are the forager groups, who are still dependent on prehistoric practice of hunting and gathering. They inhabit in the western part of the Middle and South Andaman Islands. They belong to three territorial divisions distributed in three specific geographical areas. The more Southern area is called by them as 'Boia', the northern part of the South Andaman as 'Thidang' and the middle Andaman area as 'Tanmad'. Each division is also assigned to several local groups consisting several families.

The population size of the Jarawa is only 266, as observed by the multidisciplinary research team (during 2001-2002). Generally, hunter-gatherer people (those are still in practice of hunting-gathering and have not changed their ancient economy) in the global context are of small size, consisting approximately 500-5000 individuals (Lee & Daly, 1999). Their traditional weapon is bow and arrow which are essential for their survival, hunting for food and fighting for defence.

Forest is their main source of living, They enjoy food from both aquatic and terrestrial sources. This can be known from the following table.

Table 1 : Available Food in their ecological setting.

Plant Source	*Animal Source*
Roots and tubers and seeds	Pig meat
(Wild potato, cane fruit seeds)	Pig fat
Date palm fruits	Fishes
Wild palm fruits	Crabs
Cycus fruit	Shell and molusca
Wild jack fruit	Turtle
Berry fruit	Turtle egg
Lotka*	Sea Cucumber
Gin*	Monitor lizards
Tehaigili*	Larva of wood borers
Other fruits (Green and ripe Banana, Guava,	Honey
Papiya, Coconut etc.) tender leaves of a kind	Larwa of the bee
of fern/cane	

* These are the Jarawa words.

The food from the plant sources are carbohydrate rich items, whereas food item from animal source are rich in protein substances. Both type of foods they consume are mostly boiled, roasted or smoked. It is noticed elsewhere that the plant food, especially roots, tubers and seeds are highly nutritious containing both protein and carbohydrate whenever those are cooked (Wrangham et. al., 1999). So nutritive value as well as high caloric value of those food items add strength to their body. They are healthy and energetic in general.

Generally, they take meal at least four times a day – late morning, noon, evening and night. Some take meal even in the early morning. Nature of food in different time schedule is different. They take all the food hot conspicuously direct from fire! Their principal diet is pig meat and honey. The women have restriction on some foods.

They meet their necessities from nature, both from the forest and the sea. Their needs for hut making, preparing fishing nets, other amenities, floral and shell ornaments, natural/herbal medicine etc. are fulfilled from the forest and aquatic sources. Their food resources are plenty. Their knowledge about nature is quite remarkable. They never destroy their source of collection, e.g. they never kill any pregnant pig. They use nature according to their needs.

Regarding biosocial aspect of their life, a simple picture has been observed. They have a very simple, but a beautiful system of marriage. Generally they are monogamous, but remarriage and/or widow marriage is permitted.

The marriage takes place after attainment of menarche in case of girls and in case of boys, he has an aptitude of hunting. Every individual has a selected mate, and the selection has been made by the father or any other aged relatives or aged person. The selection of mate confined within their kin group (affinal); no case of marriage among the blood relatives (consanguinal) are recorded during the present investigation.

During the study period 148 individual (i.e. 55.64% of total population) of which 81 male and 67 female are observed (Table 2).

Table 2 : Population size of the Jarawas

Area	Male	Female	Total
Tanmad	28(44.44)	35(55.56)	63(42.57)
Thidang	53(62.35)	32(37.65)	85(57.43)
Total	81(54.73)	67(45.27)	148(100.00)

0-4 years age group children are very small in number, depicting narrow base in population structure. Pedigree and reproductive history has been collected from 23 ever married Jarawa women. It is found that the range of parity is 1-8 (Table-3); never pregnant women are 5 in number, who generally are recently married, and one is pregnant. Altogether 63 birth is perpetuated by 17 ever married women, of which 71.43 per cent have survived.

Table 3 : Number of Live births of Jarawa Women

No. of Live birth	0	1	2	3	4	5	6	7	8
No. of Mother	5	4	2	1	3	4	2	–	1
Live birth	0	4	4	3	12	20	12	–	8

The average number of birth per mother is 3.71 and per married woman is 2.86. Number of dead children is 28.57 considering total mothers (Table-4). Death of female child is higher than the male child.

Table 4 : Live Birth, Surviving Child and Dead Children of the Jarawa Women

Women		Live Birth	Surv. Child	Dead Child
Never Pregnant	5	M 21(33.33)	M 17(37.38)	M 4(22.22)
Pregnant	1	F 42(66.66)	F 28(62.22)	F 14(77.78)
Pregnant at Least once	17			
Total		63(99.99)	45(100.00)	18(100.00)

Average birth rate 3.71, Average Surv. Child 2.65

Both father and mother take care of their children. Not only parents, the other members of the family or 'band' are also careful to the neonates as well as small children. The offspring are fully dependent on breast feeding till dental formation starts and continue up to 2.3 years of age along with other foods. Breast milk is given to child not only by mother, but by other women also from the same band groups whenever necessary. As they are foraging group, so during their migratory period they also carry their neonates at least from one month of age.

Over-all nutritional status of the Jarawa people is better than the mainland people. Generally, they possess good health. Prevalence of disease/morbid condition observed among them are : headache, head reeling, feeling hot, cough & cold, fever, skin disease, respiratory trouble, diarrhoea, boil etc. as well as injury due to burning, thorn bruise, fallen from tree, scorpion bite, snake bite, bee bite etc.

A good frequencies of the pneumonia, acute upper and lower respiratory tract infection are reported in other sources and diagnosed among them. During rainy and winter season they are troubled by these ailments. Low haemoglobin content is also reported among some Jarawa patients in hospital (Mathews, 2002), but Sarkar & Sahani (2002) reported a normal standard. They use both herbal and modern medicine. Even they want to be admitted in the hospital for treatments.

They are very fond of music. Singing is related with their all types of activities. Dance is also a regular practice of their life, which is related to their physical and mental health.

Conclusion

From the above discussion it can be conjectured that their food resources in their habitat are sufficient. They are very much conscious about their natural wealth and their appropriate use in wise direction. Regarding the use of forest resource, the Expert Committee's Comments (2004) after the investigation are: "The Jarawas live in harmony with nature. This is evident from their method of collection of resources. Whether it is plant or animal resource, Jarawa never destroy the source of collection. To demonstrate the point, they have cited examples. To collect bark for preparation of chest guard, they measure the exact requirement, peel off only that much bark and allow the tree to recover. Similarly, when they collect tubers, they make it sure that they collect them from mature plants only, while young plants are left undisturbed for future use. Likewise, when they have to make bow, they cut just what they need and never destroy entire tree."

Regarding population growth, it can be assessed that, inspite of limited data, the picture shows a growing population trend, due to the unlimited frequency of birth (1-8). But the frequency of child death is high. The causes of death should be investigated thoroughly. It is also observed, that in some cases, the child death is due to cough and cold, burning, crocodile bite etc. If modern medicine and neonatal as well as child care could be maintained properly by them, then the situation may be improved.

Generally, expectation of life among the hunter-gatherer population is low in the world context. If some measure can be taken against early mortality, then their population growth will undoubtedly increase. Again, if the entire forest area (Jarawa Reserve) is exclusively kept for the Jarawa people, then surely the maintenance and management of the forest will be proper.

Consequently, the existence of Jarawa population will not bear the question mark.

REFERENCES

Expert Committee (2004), Summary and Recommendation of The Expert Committee (2004). Report of The Jarawa Expert Committee. [Unpublished].

Lee, R.B. and Richard Daly (1999), The Cambridge Encyclopedia of Hunter Gatherers, Cambridge, U.K., p. 176.

Mathews, Elizabeth (2002), Health Problems and Health Care of The Jarawas in The Transition Phase. In: Jarawa Contact: Ours with them/theirs with us. Edited by K. Mukhopadhyay *et al*. pp. 155-160.

Penguin Yearbook (2005), The Penguin India Reference Year Book 2005. Edited by Derek O' Brien. p. 724.

Portman, M. (1896), Notes on the Depopulation of Melanesis, University Press, Cambridge. Cited in : The bay Islands by R.S. Mann. Institute of Social Research & Applied Anthropology, Calcutta. p. 22.

Sarkar, B.N. and R. Sahani (2002), A Demographic and Health Profile of The Jarawas of Andaman Islands. In: Jarawa Contact: Ours with them / theirs with us. Edited by Mukhopadhyay *et al*. pp. 132-147.

Wrangham, R.W. (1999), The Raw and Stolen: Cooking and the Ecology of Human Origins. Current Anthropology 40(5): 567-594.

11

THE MUNDA
A Forest Grower in Baratang

S. K. BHATTACHARYA

Introduction

The Baratang Island in the Andaman is locally referred to as only Baratang. The author of this article had the opportunity to do field work in Baratang Island in 1976, 1977, 1988 and 1990. In those days one had to go to Baratang only by boat. The Baratang Island is also known as the Ranchiwalas Island.

During the end of the nineteenth century the Britishers could surmise the potentialities of the Andaman forest. In order to extract the forest wealth in the Andamans, it was necessary for the British administrator to formulate a working plan and to give shape to that plan, they decided to bring a large number of labour force who had the expertise in extracting the hitherto untouched wealth of the Andaman forest. The Christian preachers were then well set in and around Ranchi where they could easily convert a large number of the local tribal population into Christianity. Life was not that easy for the tribals in Ranchi District due to political turmoil and associated reasons. Many of them found it difficult to collect adequate food even after moving around the jungle in quest of food throughout the day. Conversion into Christianity provided some relief to many of them. This section of people was encouraged by the church priests to go to the Andamanas where they were assured of a much better living. Initially, only a few could be persuaded to agree to cross the "Kalapani" (Andaman was called as Kalapani infamously by the people of mainland India as mainly the freedom fighters and the hardcore criminals who were awarded life imprisonment used to be deported to the Andamans.)

History of Migration

Those who ultimately reached Andaman could detect the new place as heaven for them and not as the dreaded "Kalapani". Good food, good shelter, good working condition and good environment made them happy in their newly found home in the Andamans. So, batches after batches started arriving at Port Blair. The British administrator also started drafting plans to explore more and more areas to extract valuable forest produce and the importance of these sons of nature gained momentum. These well behaved and quiet types of forest-dwellers of the Ranchi district soon became the most dependable labour force of the Britishers in the Andamans. They were soon taken to Nicobar groups of Island also. Many of them were also given jobs in the Department of Marine. So the Chhotanagpur tribals are seen almost at every nook and corner of the Andaman and Nicobar Islands.

The Second World War began and, for a brief period, the Andaman came under the Japanese occupation. Many of these tribals left the Andamans during this period and those who stayed back had to undergo torture in the hands of the Japanese Army. Soon after the Indian Independence there was flow of these people going to the Andamans again. Among them, there were many who had worked in the Andaman before and during the Second World War. This time they made their rightful claim for the permanent settlement in the Andamans. After a patient waiting for long, some of them were awarded land to settle in the Baratang Island, which was almost a barren field in 1959. Each family of these jungle dwellers got 5 acres of lands to settle in the barren field. Within a period of about 20 years these people made the Baratang Island full of green.

The history of migration of these Chhotanagpur tribals to the Andamans was collected from a group of elderly persons during field work in Baratang Island in 1976 and the data in this regard were checked time and again from the Chhotanagpur tribals working in Chatham Saw Mill and Dairy Farm Village. In 1990 the Baratang Island was no more a barren land. The people settled in this island made paddy field, vegetable gardens, which were found to be surrounded by big trees. These settlers expressed their happiness in their present living condition. Some of them, however, had regular contact with the counterparts left behind in the mainland India but not a single individual was found eager to go back to their original homeland for good. They are found making a new forest for themselves and more so for their progenies.

Nature of Data

The above part of this write-up is based on the socio-economic data collected from them as well as on the observation made on the surroundings in the Baratang Island during the author's fields works under a project titled "Microvariation among the Chhotanagpur tribals settled in the Andamans". The data on Anthropometry, Serology, finger Dermatoglyphics and Demographic Parameters were collected from the Oraon, the Munda and the Dudha Kharia of the Andamans during 1976, 1977 and 1978.

Health is product of body and mind. Supply of adequate food, a good shelter and a peaceful living condition have a positive impact on the growth of the body. The new settlers of the Baratang Island, it seems, know these things well. So they keep the surroundings clean. Every family gets safe drinking water. Availability of food to these people has already been discussed. All these factors put together have brought them favourable condition to keep good health, which is a must for the road to development.

So it is important to measure the effect of good environment and availability of adequate nutritious food to make a comprehensive report on a population. In this paper only the data collected from the Munda of the Andamans have been used to measure the development based on some demographic and anthropometric parameters.

Biological Parameters

The average live birth among the present sample of the Munda is 6.22 and the average number of surviving children per ever-pregnant Munda Mother is 4.89. The demographic parameters have already been analysed in details (Bhattacharya, 1995). It will be more repetition to put that elaborated work here again. But the result obtained from that study is required here to assess the development among the Munda of the Andamans. One sentence from the above study will speak for it self. The sentence is that "The Munda in the Andamans, like the Oraon, will grow at a considerably higher rate in the near future", (p. 70).

In another attempt (Bhattacharya & Bhattacharya 1986) five biometric characters namely, calf girth, upper arm girth, chest girth and stature in addition to body weight were used to assess the healthy components of the Munda of the Andamans. From the data on those parameters along with robusticity index of pigment suggest that the Munda of the Andamans have a sound physique. The observation made by Bhattacharya and Bhattacharya (1986) among 25 populations in this respect, is that "only the Nicobarese groups studied by Ganguly, 1964 show better body built than the Munda of the Andamans."

They get deer, wild boar, honey and many other edible things from the forest they have already grown in the Island. This has substantially contributed to them to achieve this feat in respect of their health.

Man has spent the largest period since the beginning of the human evolution as a forager. The quality of the life of the foragers speaks for the success of the human growth. The genuine forest dwellers never destroy their forest. This paper shares the experience of forest dwellers growing forest in almost a barren land given to them as gift for their extremely laborious work in the Andamans for generations.

What the Munda of the Andamans and their neighbours both in Ranchi and in these Islands are doing is in their culture. They have learnt the technique to use the forest for their survival and protect their surrounding nature also for their survival. This paper is a learning for the author to understand that the protection of forest will be the best in the safe hands of the forest dwellers. Who other than the forest dwellers, mostly the tribes, understands it better?

REFERENCES

Bhattacharya, S.K. and Sarmistha Bhattacharya, *The Physique of the Oraon the Mundas and the Dudh Kharias in the Andamans in Hum. Sc.* vol. 35, No. 4, 1986.

Bhattacharyya S.K., *Demographic Variable among the Munda: A Migrant Population in the Andamans in J. Anthropological Survey of India*, 44, 1995.

Dash Sharma, P., *Physical Anthropological Studies on Mundas, Oraons and Dusadhs of Bihar*, 1978.

Ganguly, Pranab, *Physical Anthropology of the Nicobarese*, Anthropological Survey of India, Calcutta, 1976.

Vidyarthi, L.P., The Ranchi Tribals in Andaman and Nicobar Islands in *J. Soc. Res.* Ranchi.14:50-59, 1971.

12

FOREST AS THE MATERIAL BASIS OF TRIBAL SOCIETY DURING COLONIAL PERIOD

ASHOK KUMAR VERMA

Introduction

Forests have figured as one of the potential natural resources of a country. Tribals and forests have been interdependent and interrelated. They have developed their culture and style of life according to the forest because forests have been fulfilling their various needs since times immemorial. The forest has maintained the very existence of the tribals since centuries. Tribal culture and forests have been inseparable entities. But their lordship on forests and forest produce did not remain safe with the passage of time.

During the ancient period the entire area of Chhotanagpur and Santhal Pargana was full of dense forests. In the Vedic and epic period the tribals enjoyed full rights and lordship of the forests. Within the span of the Mauryan period rights over forests were not disturbed. Even today, the tribals are fond of living in and around the forest. They are utilizing the forests for their day-to-day needs. Forests are fulfilling the social, economic, cultural, religious, political, health and medical needs of the tribals, moving from pasture to pasture, opening out the country along fertile valleys and streams.

Division of Chhotanagpur Forest

The first geographical and statistical report of the Chhotanagpur division under the topography survey department, published by

Capt. 'Depree' in 1868, states that the division comprised an area of 43,722 sq. miles. The chief forested parts are the slopes of western parts and the bordering ranges in the north. The Bishunpur thana and the long range of hills which runs north from Palkot to the borders of Palamu are generally covered with forests. The Chhotanagpur forest alone covers an area of 1830 sq. miles i.e. 66% of the total forest and the remaining 936 sq miles belong mostly to the Santhal Pargana.

In Chhotanagpur the places with maximum forest covered regions were – Hazaribagh (48.36%), Palamu (44.68%), Giridih (34.03%), Lohardaga (29.06%), Singhbhum (28.02%), Gumla (25.53%), Ranchi (20.15%), Sahebganj (12.26%), Dumka (10.75%), Godda (10.32%), Deoghar (8.94%) and the lowest in Dhanbad (7.89%).

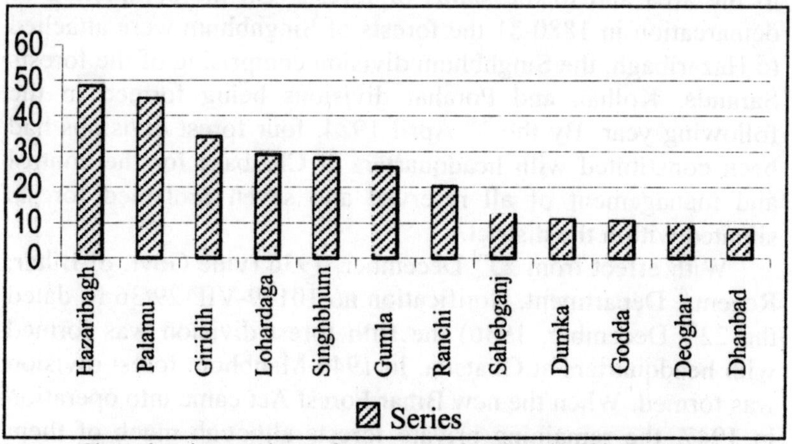

Chart showing maximum forest covered regions in Chhotanagpur

Reckless destruction of the forests in Chhotanagpur began from the time of Akbar and continued upto the first decade of the 19th century, when the British had already established their sovereignty in this land. No serious action seems to have been taken for the preservation of Chhotanagpur forest until the report of J.F.K. Hewitt, the Commissioner of Chhotanagpur division (1882-85) and the subsequent report of Dr. Schlick (1885) reached the Central Government, when considerable damage to forests had already been done.

The first investigation into the question of reserving the forests in Chhotanagpur was sponsored by Dr. Anderson, the conservator of forest in 1864. The forests of Singhbhum district were examined by Captain Losack, Deputy Conservator of forests in 1870-71 and again by the Davis, Deputy Conservator of forests in 1879-80. By notification dated the 10th of August, 1875 (under Act VII of 1865) Saranda Pir was declared Reserved forest. An area of 1,99,740 acres was notified under section 4 on the 26th November, 1880 and finally declared reserved under section 19 on the 17th May, 1882 with effect from the first April, 1882.

The history of protected forest dates back to 1903-1905 when 17 blocks were declared protected. Subsequent to that there were various additions and modifications both with regard to the area and to the status of forests. On the completion of demarcation in 1880-81 the forests of Singhbhum were attached to Hazaribagh, the Singhbhum division comprising of the forests Saranda, Kolhan and Porahat divisions being formed in the following year. By the 1st April 1924, four forest divisions had been constituted with headquarters at Chaibasa for the control and management of all reserved and seven protected forests situated within the district.

With effect from 23rd December, 1936 (vide Govt. of Bihar, Revenue Department, Notification no 10149-VIF-29/36 R, dated the 22nd December, 1936) the fifth forest division was formed with headquarters at Ghatsila. In 1946 Manbhum forest division was formed. When the new Bihar Forest Act came into operation in 1947, the remaining private forests although much of them had been devastated by then, were also notified and finally taken over for management by the Government.

After the merger of Saraikela and Kharsawan states with the state of Bihar, the forests were formed into two separate ranges of the same name, now a part of Dhalbhum division. After the abolition of Zamindari, the rights of management of private forests of the district were vested in the state of Bihar under Forest Act. There were only two types of forests namely Protected and Reserved and other classifications were superseded. Local forests were marked Reserved and subsequently plundered for commercial purposes.

In 1894, the British Government formulated forest policy, which aimed at bringing the forests under state control. Forests became the main tool of administration in British India. The British had an aim of gaining in their imperialist attitude through forest policies and forest laws. In order to suppress the revolution of Birsa Munda, the British were in need of faster means of communications and the only way was expansion of Railways. To lay Railway tracks swiftly, the British attacked the forests.

In 1910, O'Malley wrote in the Singhbhum District Gazetteer that as soon as the running of goods train began on the Bengal-Nagpur line the market of forests boomed. Between 1895-98, 20,921 trees were cut down to get 8,07,627 broad gauge sleepers. The revenue of 8.50 lakhs was obtained from it. Within five years the revenue increased in such a way that in 1903-04 it was 8,40,000 rupees and in 1907-08 it increased up to 1 lakh 90 thousand .

The British rulers gradually transformed this 'loot' into 'law'. Special division of forests were branded Reserved forest. The rest, large and dense forests were termed, as Protected forests. Both types of forests came under the private control of Imperial Forest Department.

After the suppression of Birsa Munda revolt, the British rulers drove away the tribals from Khuntkatti villages coming in the forest areas, governed by the forest laws. On the basis of documents, Father Mathew Areeparampil observed in this regard that protected forests were marked in Porahat region in 1890. The revenue from the forests went to the king .

Then came the Chhotanagpur Tenancy Act in 1908 in which the British rulers pretended to give special rights to Mundari Khunkattidars over their land . It is sufficient to observe here that the Act was a mere eyewash. Regarding decision in 1910, *J.M. Arden* observes in this connection that seven villages were made forest villages. For this the civil departments had released Rs. 1.931 from the forest department.

Role of Forest in the Life of Tribals

Forests are an important adjunct to tribal culture and economy. The life and economy of the tribals is mainly based on natural resources comprising the primitive methods of agriculture and collection of forest products. The forest has played a significant role in shaping the socio-economic, religious and cultural systems of tribal societies.

Role in Social Life

The forests are such an integral part of the tribal society that tribals also believe that the spirits of their ancestors reside at different places in the forest. They worship bushes, trees, animals, birds and sarana groves. They make offerings of flowers, leaves etc. to these entities, considered sacred by them to please their gods and goddesses.

Since times immemorial hunting too has been an economic as well as social activity of the tribal community. They used to hunt different types of animals and birds. The horns, nails, bones of the animals hunted were used to perform magic. Ornaments and handicrafts were prepared from its horns and bones. Annual Ritual Hunt was an annual religious exercise among the tribals. Worship was an important aspect of the exercise. There was a great deal of co-operation among the tribal groups during the annual ritual hunting, which has a great cultural relevance.

Just like *Bitlah*a (a tradition of imposing ban/fine on those involved in extra marital affairs) a similar system of rules and regulations existed in annual ritual hunt. Annual ritual hunt is known as *Sendra* and there were various taboos and traditions that were the same. The logic behind this practice is return of debt to the forests. Sendra was observed in the localised region. But there have been several threats to the practice due to deforestation and increasing population. The withdrawal of tribal rights over forests and its subsequent conversion to reserved and protected forests has played a greater role in the degradation of the forest. The Government has banned hunting wild animals and birds now-a-days. There have been several ideas floated in the past to ban sendra. An alternative or modification is necessary instead of banning it completely, which would otherwise result into death of a tradition.

There were several tribes in Jharkhand in specific totems representing animal forms, which were never harmed. Such hunting practices were carried out in groups and in specific times of the year.

If Annual ritual hunt is banned, *Dongse* a particular tribal war dance involving Santhal men will lose its relevance. In addition *Pitkal Dance*, a dance form depicting a man dancing with wild animals along with hunting implements would also be lost. There are certain restrictions regarding the practice: elephants, tigers and bears are not killed. Also, totemic animals are not killed, thus ensuring that a spectrum of animals that are ecologically very important remain untouched. The practice could be modified into a sports festival where only few hunters could be sent for the hunting of previously marked animals. This would check the numbers of animals killed and also ensure that the practice is not obliterated.

Hunting festival is carried out in Dalma, Naranbeda, Buda budi, Makna, Vankari, Araburu, Dobraburu, Ayodhya etc. All tribal communities like the Munda, Santhal, Ho, Bhumij, Oraon, Kharia take part in the exercise. Almost 60 per cent of the village male population between the age group of 12 to 50 years carry out the hunt. Annual ritual hunt practice is a traditional practice among the tribal communities and must be conserved at all costs.

Role of Forest in the Economic Life of Tribals

The economy of tribals is still dependent upon the forest. They construct their huts and houses from bamboo, wood, leaves etc available in the forests. The collection of minor food produce is still the main economic activity of majority of tribals. They collect roots, shoots, flowers, fruits, seeds, vegetables etc. from the forest in different season. They collect different forest products, honey, wax, lac etc. to earn money. They earn wages as forest labour. They are dependent upon the forest for the medicine and collect medicinal herbs from the forest. Different factors of economic developments such as production, consumption, distribution etc. are closely related to the forest products. The tribal economy and the forests are closely inter-related. We can discuss it under the following heads — Hunting, gathering of food, fuel collection, grazing of cattle and collection of fodder, shifting cultivation, forest labour etc.

Food Gathering

Since pre-historic times food gathering has been an important economic activity. Tribals engaged in pastoral agriculture, artisan and other kinds of economic activities can also be treated as food gatherers, gathering food in and around the forest. They also have knowledge of edible roots found below the soil. They collect the shoots of the creeper plants Bathua, Gandhari, Kandas, etc from the forest. They also collect various leaves to use as food items i.e. Pipal, Bargad, Gular, Pudina, Tamarind etc. The tribals use edible flowers of Mahua, Karam, Arjun, Jamun etc as vegetables. The ripe fruits of mango, bargad, gular, jamun etc are used by the tribals both as sources of food as well as income. Besides, they prepare edible oil from Mahua seeds and also gather mushrooms from the forest. Other important items collected by the tribals are honey and honeycombs. Both mushroom and honey are also sold by them in the market. Thus, the collection of minor food produce provides the occasion of gainful engagement to the tribals round the year.

Minor Forest Produce

The minor food produce forms an important part of the tribal economy. Tribals has enjoyed the right of collection of minor food produce by tradition. In the beginning they used to do the collection of MFP only for the consumption of the reciprocity. Gradually, they began to barter the MFP with the neighbouring non-tribal communities. Tribal areas in Chhotanagpur and Santhal Parganas region are potentially rich in MFP items like Kendu leaf, Sal seeds, Mahua, Kusum, Karanj, Palas, Harra, Bahera, Amla, Neem, Honey, Wax, Lac , Gum etc. Major parts of the MFPs are sold from tribal areas in raw form. MFPs are classified by the national commission on agriculture as fibers and flosses, Grasses, Bamboo, Seeds and Canes, essential oils, oil seeds, Gum, resin, drugs spices, leaves, lac, tussar, Silk, Wax, Honey and other products. Sabai grass is a most useful commodity; it serves household purposes but it is exported mostly for the manufacture of paper. It is also utilized locally for string making. Traditionally, tribes of our country were dependent on the collection of MFP for sustenance as well as supplementary income. The non-tribal traders introduced barter system in the tribal areas. They supplied essential commodities to the tribals in exchange of forest produce and agricultural produce. *Hats* or weekly markets have been part and parcel of the tribal people. In the year 1848 the first weekly market was established at Chaibasa. In 1887 this number increased to twenty six and then to thirty seven in 1920. Salt sellers, ornament sellers, cloth merchants etc. came to these 'hats' with their wares. They exchanged their wares for grains, oil seeds and agricultural products with the tribals. Economic transaction takes place in Hat. Hat became the centre of socio-cultural activities. Tribals collect dry branches of the tree for the purpose of fuel and make bundle of firewood for sale. They get raw materials for making barks, rope and rope model articles etc. from the forest.

Medicinal Plants

Tribals get medicinal plants from the forest. In the forest of Chhotanagpur region immense medicinal plants are available such as Amla, Harre, Bahera, Fruit and seeds bark etc. They collect forest produce of medicinal value and get collection price as wages. Satwar, Papuravi, Amarlata, Neem, Bel etc are important herbs of medicinal value. The flowers of Mahua, Jamuns, Palas etc are used as medicines. The oil of Mahua, Karanj, etc is also used as medicines.

Lac and Tussar

There are many other important forest products in this region like Lac and Tussar. The production of Lac in this division is more than half the output of the whole world. But this industry of Chhotanagpur is at present greatly depressed due to advent of cheap synthetic stuff from abroad, resulting in the decline of its prices. From the epic of Mahabharat and the *Atharvaveda*, there has been reference of lac. The lac industry provides opportunity for socio-economic development of tribals. It is a natural resin of animal origin. The important trees on which the lac insects breed and thrive well are Kusum, Palas, Ber, Babul, Khair, Arhar. It is used in making toys, sealing wax, gramophone record, printing ink, filling ornaments, silvering the back of mirrors etc.. The important assembly centres of lac were Manoharpur, Goilkera, Chaibasa, Jhinkpani, Hatgamaria, Bharbharia, Jayantgarh and Barkela. An annual production of all kinds of lac was estimated to be at 30,000 and 35,000 mounds between the years 1931-1956.

The Indian council of agricultural research has established Lac Research Institute at Namkum in Ranchi. Manbhum and Palamu were the maximum production regions of Lac in Jharkhand.

Tussar industry in Chhotanagpur is one of the oldest industries. It was carried on by the Hos as a subsidiary occupation. Asan, Arjun, Sal, are food plants of silk worm. Tribal people living in and around the forests were attached to this industry. It provided opportunities and employment with minimum investment.

During the Craven's settlement the Asan trees in the Kolhan stood at 14,54,136. Between the years 1899 and 1916 the Tussar industry suffered a set back. It began with the levying of eight annas for every 100 cocoons in reserved Government forests. Then came the long drawn drought of 1915-1916.

The Government took a remedial measure and established a silk rearing farm one mile away from Chaibasa railway station in 1906 on area of 55 acres. The Government hoped to check the decline of the industry but the number of cultivators did not increase between 1906 and 1916. In 1936 the Government took another step and established the Tussar Seed Supply and Research Station. The useful work done by the Research Station revived the industry to a certain extent.

Biri Industry

The Kendu leaves from the forests were used in this industry. It gave employment to about 30 to 40 thousand labourers. The Chhotanagpur forests grew these leaves abundantly.

Hence the tribals and forests have been interdependent and interrelated. As the forest are gradually disappearing, economy based on the forest also tends to become rarer. But still the forest provides the tribals with some important articles namely wooden poles for building, and agricultural implements, leaves for making leaf cups and plates.

Conclusion

Thus the forests have a very important role to play in the development of tribals but it has not been so due to the degradation of forest. Following are the most important factors for forest degradation:-

- Lease of Forest by the kings for clear felling purpose.
- Excessive quarrying activities in the forest.
- Developing infrastructure, laying concrete roads and rail routes.
- Lax vigil by the forest department.
- Creation of forest development corporation.
- Extraction of root stock.
- Coup permit to contractors for cutting forests.

The impact of these forest degradation on tribals can be stated as follows :-

- Scarcity of essential forest products for need fulfilment.
- No fuel wood – an essential element of village life.
- Excessive biotic pressure – Grazing.
- Scant and irregular rainfall.
- Complete loss of livelihood specially for the tribal community.

Forests should be seen as a rich and viable source to meet the developmental needs of the tribals. In spite of the good intentions of the National Forest Policy, 1988 and that of 1990, it has not been possible to meet the expectations of the tribals. Indian Forest Act, 1927 continues in the present day without any significant changes in its provision. In order to formulate a suitable policy for effective management of forests, there is a need to bring the corresponding changes in related Act and policies. In Chhoṭanagpur, there is still an extensive scope for the development of the forest. Unless uniform development policies for the tribal communities are formulated and implemented, it will be difficult to achieve the desired goal of tribal and forest development. The sincere implementation of Government policies can make the forest much more beneficial for the tribals from the economic point of view. They should be given freedom to do the collection of MFP in the forests. They should be given facility for the collection of dry woods, bushes, branches, leaves etc. for use as fuel. They should be given right to collect fodder and to do herding in the forest. They should be given right to do the collection of medicinal plants. Forest policies should be framed seeking the opinions of the tribals and villagers. The schemes of forest development and protection should be formulated in village level committees. Exploitation of tribals by the Government agencies should be treated seriously. The concept of sustainable development should be adopted in the formulation of forest policies.

REFERENCES

B.C. Sharma and Kesri Vikram, Chhotanagpur Ka Bhugol, 1997.

L.N. Bhagat, U.N. Verma, Tribal Area development – Need for Alternative Strategy, 2000.

L.S.S.O' Malley, Bengal District Gazetteer, Vol. xx, Singhbhum, Saraikella and Kharsawan, Calcutta, 1910.

M. Sahu, The Kolhan under the British Rule, Calcutta, 1985.

Mangobinda Banerjee, An historical outline of Pre-British Chhotanagpur, 1993.

Mathew Areeparampil, Documents, 1988.

P.C. Roy Choudhury, Bihar District Gazeteer, Singhbhum, Secretariat Press, Patna, 1958.

Paper Proceeding of Annual Ritual Hunt, SNTI, Jamshedpur 10th Sept, 2003 Organised by Indian Institute of Forest Management, Nehru Nagar, Bhopal.

Sources Field Studies in Eastern India.

Vijay Sankar Upadhyay, G. Pandey – Tribal Development in India – A Critical Appraisal, Crown Publication, Ranchi, 2003.

13

SYMBIOTIC RELATIONSHIP OF TRIBALS AND FOREST
Role of Tribals in MFP Marketing in Eastern India

ARUNA SHRIVASTAVA

Introduction

There has been a close relation between forest and tribals since times immemorial. Whenever the term tribe is used, an image of forest dwellers comes in mind. The forest has maintained the existence of the tribals since centuries. On the other hand, tribals have protected the forest since long. In this way, there has been a symbiotic relation between forest and tribals. So long as the forest was under the possession of tribals, there was no decrease in the forest areas. But as soon as the tribals were made forest labourers from the forest lord, decrease in forest areas started taking place. It has placed a question mark on the survival of many tribals whose primary means of stay is the collection of MFP (Minor Forest Produce) even today. Tribal culture and forests have been interdependent and inter-related. The forest has played a significant role in shaping the social, economic, religious, political and cultural systems of tribal societies.

This paper hence tries to focus light on the role of tribals in the collection of MFP throughout the year. It is found that the collection of MFP is still the main economic activity of the majority of tribals in India. The paper has delimited the area to eastern India with special reference to Jharkhand to emphasise this economic activity.

The forest and tribals are interdependent. They are inseparable. They cannot exist in isolation. The existence of one cannot be thought of in absence of the other. The paper also emphasises the symbiotic relationship through economy and role of tribals in marketing of the MFP.

To understand the symbiotic relationship of the tribals and forest, we must consider the tribal economy and the forest economy.

Forest Economy and Tribal Economy

The system by which human resources and natural resources are governed is known as economy. The tribal economy is dependent on forest because economic institutions like production, consumption and distribution, market, trade and labour etc are related to forest. This relation of the forest and tribal economy can be classified in to the following heading:

1. Hunting
2. Food gathering
3. Collection of MFP
4. Collection of fuel
5. Pastoralism and collection of fodder
6. Artisan activities
7. Shifting cultivation
8. Forest labour

The hunting was such an important cultural feature of tribal society as even today one finds the existence of hunting tools among the tribals. The hunting implements of tribals include net, spear, bow, arrows, gulel etc. The Government has banned the hunting of wild animals and birds. They used to hunt birds and animals like deer, hare, rabbit, pigeon, panduk, titar, peacock and sparrow etc. Hanuman shikar is one of the special features of the Birhor tribe of Jharkhand.

The ban of the Government on hunting has brought adverse effect upon the festival of Sendra and Janishikar in which hunting was organized by the tribal women in groups. Previously, they used to trap animals and birds for sacrifice on religious and festive occasions. Now they rear goat, hen, cock and duck for the purpose of sacrifice. Those who do not rear purchase the sacrificial animal or bird from the market.

In fact hunting and food-gathering together represent the early stage of economy of mankind. The example of food gathering among the tribals can be seen in every part of the country. Tribals are also engaged in other kinds of economic activities such as pastoralism, agriculture, artisan and are also doing food gathering in and around forest. Food gathering is an economic reality for tribals of different cultural types. The tribals are collecting edible roots, leaves, flowers, fruits, seeds, mushrooms, honey, wax and sag (leafy vegetables).

Nomenclature of Minor Forest Produce (MFP) and British Government

During the British period, the forest produce was divided into two categories.

(a) Major Forest Produce (b) MFP

Timber yielding trees, ornamental trees and trees used for paper pulp came under the category of Major Forest Produce. The British Government had put a ban on felling of those trees which came under the Major Forest Produce by the tribals and they themselves were interested in cutting the trees for their own purposes like preparing ships, furnitures, making buildings and pulp and paper industries. They left the tribals for collection of root, leaves, flowers, honey, wax, lac etc. and they considered these produce of no value. They used the term "MFP." The British rarely collected the MFP but those forest products were of great value for the tribals from the point of view of subsistence since times immemorial. Tribals have enjoyed the right of collection of the MFP by tradition. This tradition also indicates the symbiotic relationship since prehistoric times.

In the beginning, tribals used to collect the MFP only for consumption and reciprocity. Then they began to barter the MFP with the neighbouring non-tribal communities. When local traders made the tribals aware of the commercial value of some of MFP, the tribals began to do the collection of MFP to get some cash by selling MFP in the hands of the local traders, banias and mahajans. In this way the MFPs have become the substance of marketing and earning cash.

Categories of MFP and the Value of MFP during Independence and Post Independence Period

The MFPs are available in the forest in different seasons of the year and tribals are engaged through-out the year in collection of MFP. 74 Tribal communities who come under the PTG (Primitive Tribal Group) are dependent on collection of MFP.

The MFP can be divided into following categories:-

(a) Plant yielding essential oil.
(b) Medicinal plants.
(c) Fatty acids.
(d) Gums and resins.
(e) Dyes.
(f) Fibre and Flosses.
(g) Bamboo and cane.
(h) Plants yielding edible roots, shoots, flowers, fruits, seeds etc.
(i) Honey and wax.
(j) Fodder and grasses.
(k) Tendu leaves for Biri making.
(l) Mahulayan leaves for Plate making.
(m) Lac.
(n) Leaves for making Dona (cups).
(o) Chop or bark for making ropes.

Before independence, the value of MFP was so low that the forest dept did not regard them as important sources of revenue. There are many items which have acquired commercial value after independence nationally and internationally. Thus it appears clearly that the nomenclature MFP is not suitable for these forest produce.

In Madhya Pradesh alone, collection of Tendu leaves ranges from 44 lac standard bags to 55 lac standard bags. Madhya Pradesh Forest Department earns Rs. 400 to Rs. 500 crores annually from the collection of MFP. At the country level, revenue from Tendu leaves is much higher because other states like Orissa, Jharkhand, and U.P. are also engaged in revenue earning from the Tendu leaves. There are many other items which have acquired commercial value nationally and internationally. Their turnover has risen after independence.

Today, the tribals of our country are at different levels of economic development. This has affected their dependence on the collection of MFP. The Primitive Tribal Groups (PTG) are still dependent upon MFP for their survival and livelihood. In Jharkhand state out of 32 tribal groups eight are PTGS which are still dependent upon MFP for their survivals and livelihood. Many of them are still in the food-gathering stage. Other tribals have been collecting MFP only as a matter of traditional rights with a view to earning supplementary income. According to some estimates, nearly 80 to 85 per cent tribals in our country are still engaged in the collection of MFP in one way or the other.

A study conducted by Administrative Staff College, Hyderabad (1978) revealed that between 10 to 55 per cent of income of tribal families in the major tribal concentration states of M.P., Orissa, Jharkhand and Andhra Pradesh was based on the collection and sale of MFP.

Private Traders' Entry into the Marketing of MFP

Historically, the tribals were collecting the MFP for their own consumption and use. A change in this system took place after the in-flux of non-tribal traders into the tribal areas. The non-tribals introduced barter system in the tribal areas. These traders initially supplied essential commodities like salt, tobacco, kerosene oil, spices etc. to the tribals in exchange for the forest produce and agricultural produce. This system introduced a spirit of competitions among tribals to collect more and more forest produce and get more and more essential commodities in exchange. Realizing the commercial value of the various MFP, the traders introduced cash in their transaction with the tribal people. The introduction of cash brought more competition among the tribals for collection of MFP. The private traders used to pay very low price for the produce. Sometimes they used to pay loan in cash against the MFP. If it is not possible to repay all loan from the sale of MFP, the traders kept the land of the tribals in mortgage and those who did not hold the land had to work as bonded labourers to the fields of the traders or mahajans. These mahajans had full control over the Haats of tribals. These mahajans were very clever. They used to exploit the sentiments of the tribals by distributing Biri, Cigarette, Tobacco etc. so that they were able to buy the MFP at the reduced rates.

During the British period the tribals organised themselves against the exploitation of the leaders in the form of revolts. But the British Government suppressed the revolts in the name of law and order.

In the post independent period the then Madras Government was the first to appoint a committee in 1952 to examine the revenue for improving the socio-economic conditions of the tribals. This committee is known as Malayalam Committee. The Committee in its report mentioned that the money-lenders (mahajans) were responsible for the misery of the tribals. The socio-economic advancement is not possible without the elimination of the middleman.

In the 5th Plan period LAMPS were started as ground level Government Agencies to purchase MFP collected by tribals. But various study reports reveal that private traders still dominate the purchase as marketing of the MFP collected by tribals. Such type of practice is found in Chhattisgarh, M.P., Jharkhand and West Bengal even today.

The MFP fall under two categories:

(a) Those, which are raised in the forests like lac and tassar.
(b) Those, which grow in the forests naturally.

Most of the items come under the second category. These are again divided into two as nationalised and non-nationalised. With the collection of nationalised the private traders do not come into the picture. The collection is organised by State Forest Departments either directly under their supervision or by engaging LAMPS, where the LAMPS are weak by private traders. Collection is made even through State Forest Development Corporations and Tribal/Development Co-operative Corpora-tions of LAMP, Large Area multipurpose co-operative societies, TRIFFD–Tribal Co-operative Marketing Development Fede-ration of India Ltd, **MFP** Trade and Development Federations.

Government Agencies Associated with the Marketing of MFP

Three Government agencies at present are associated with the marketing of MFP which are as follows:

(1) LAMPS – Field level collection agency.
(2) State Forest Development Corporation and Tribal Development Co-operative Corporations/MFP trade and Development Federation-state level agency.
(3) TRIFED National and International Level Agency.

Marketing of MFP

Marketing of MFP items exists both within the country and outside the country. There are only a few items of MFP like gum, karega, lac etc. which have very little domestic consumption (10% to 20%) 80 per cent to 90 per cent of these items are exported to different countries.

The marketing strategy of these items depends upon the demand and supply situation in the international market.

Until 1987 National Level Apex Organization TRIFED was set up for planning and developing market strategies of different items of MFP with the view that tribals receive the benefits of price in the terminal market. The marketing of MFP collected by the tribals was dependent on the private traders in the secondary and terminal markets. Most of the MFP items collected by State Federation/Corporation are disposed off by involving traders or through traders cum-auction. Very rarely, the state agencies have gone in for direct negotiations with the traders from the secondary markets. In most of the cases, it is limited to the state and, at times, even to the district/districts in which MFP collections are stored. This arrangement works to the advantage of the private traders located in near by secondary markets. The tender price is finalized by the state agencies themselves except in case of sal seeds in Orissa where a tender committee has been set up at the state level. But in Orissa, the experience has been that it is a time-consuming and lengthy process of going through a state level committee.

Different items of MFP have by now established secondary market which work as feeder for the terminal market. Before the advent of TRIFED very few State agencies established direct marketing relationship with the terminal users or markets. They have preferred to follow that even now, except in cases where TRIFED offers a much higher price.

The study of some secondary markets reveals that the traders of the secondary markets generally operate as commission agents for the big traders of terminal markets. These traders in terminal markets control the interland supply as well as meet the export requirements. There is an excellent nexus between the operators in the secondary markets and those in the terminal markets with many in the latter employing all possible means to reach down even to the primary level of collection in case of non-nationalised items through their own hired agents. The main market functionaries are identified as "Paikars", who operate as procurement agents of various principals, namely the Lac factory owners– Kothiwals, Rakhiwals, Arhiatias and Bhattawals. These paikars are purely commission agents.

TRIFED has been taking interest and initiative every year and announcing a much higher price to Lac stick procured from the tribals. But it has not been able to make any impression so far due to lack of support from the state level agencies which are also its constituents for formulating a comprehensive policy on Lac marketing, so that tribals back home could be given full benefits of the international price. But the Central Government has not made any decision as yet and the marketing of Lac continued to be in the hands of few powerful vested groups.

Conclusion

In this way there has been a close relation between the tribals and forest. Tribal culture and forests have been inseparable identities. Both are interdependent and inter-related. In this way there has been a symbiotic relationship between the two. So if Government is willing to protect the forest areas, they have to make policies accordingly because the existence of one cannot be thought of in absence of the other. They exist symbiotically. The collection of MFP is still the main economic activity of majority of tribals. Some of the PTGS of Jharkhand are still surviving on collection of MFP. Each season provides the tribals an occasion for the collection of varieties of M.F.P. Thus forests provide gainful engagement to the tribals round the year. They collect roots, shoots, barks, leaves, flowers, fruits, seeds, vegetables etc. from the forest in different seasons. They eat them raw or cooked. They collect grasses and bark for the preparation of rope, green grasses and leaves to use as fodder and dry leaves and wood to use as fuel. They collect honey, wax, lac etc. to earn money. They also exchange and sell some MFP in the local haats. To get cash from the local haats these MFP, through the paikars, go to secondary markets and finally to the terminal markets. Through terminal markets some of the MFP are exported to the international markets.

Thus the symbiotic life of tribals with forest plays a very important role in the national economy indirectly. Hence Government should take a strong decision and formulate policies in marketing the MFP directly from the tribals so that they are benefited and their economic condition improves. It will play a vital role in the development of states such as Jharkhand which is highly populated by the tribals.

REFERECENCES

Deogaonkar, S.G.– *Tribal Development Plans*, Concept Publishing Company, New Delhi, 1992.

Deogaonker, S.G.– *Tribal Administration and Development with Ethnographic Profiles of Selected Tribes*, Concept Publishing Company, New Delhi, 1994.

Haimendorf, Furer, C.Von, Tribal Population of Hyderabad, in Upadhyay V.S. and Pandey G, (ed). *Tribal Development in India*, Crown Publication, Ranchi, 1945.

Oraon, P.C., Land and People of Jharkhand, *A JTWRI Publication*, Welfare Dept., Government of Jharkhand, Morabadi, Ranchi, 2003.

Upadhyay, V.S. and Pandey G. *Tribal Development in India: A critical Appraisal*, Crown Publication, Ranchi, 2003.

14

FOREST AND DEVELOPMENT
Anthropological Perspective

DIPAK PIPLAI

Introduction

In Anthropology, the importance of forest is not merely a question related only to soil erosion, environment pollution, ecological imbalance etc., but as a mother of many human societies. Forests provide the living-space to a huge number of people (though exactly how many of the 1028 million people in general and 84 million so-called 'tribal', i.e. the aboriginal/ adivasi people [2001 census] of India live in forest areas, is not known); it is the basis of their social, economic and cultural life; it is a pivot of their emotions, ethos and aspirations. Moreover, forests also affect the people's life of distant places in many ways. Hence, forest demands serious attention of Anthropology.

Forest *versus* People

Food and Agricultural Organisation (FAO) defined 'forest' as a "land with a tree canopy cover of more than 10 per cent and area of more than 0.5 hac." Apart from the presence of trees, forests are also determined by the absence of other predominant land uses (Penguin, 2005). When forests provide shelter and job, food and medicine, defensive and recreational materials, coverings and ornaments to its sons, human society then should also help the forests, with a deep feeling of gratitude, to thrive on earth. 'Civilised' society should not look upon the forests only as a means of 'revenue earning' and making the state policy accordingly; but should consider the forest truly as a protector of human life.

93.8 per cent of the total Adivasi population of India lives in and around the forests. ''Forefathers of these tribals settled themselves within the forest and had been surviving on forests and other available natural resources in the forest. As a result, their economic, social and cultural practices closely revolve round the forest life-cycle and they become the foster children of the forest'' (Basu, 1987). ''The totality of the tribal culture revolves around the forest.... there has been an in-built concern about nature and its importance in the tribal customs and traditions'' (Deogaonkar, 2004). So, the concern of modernity should be the 'Forest and People', not the 'Forest *versus* People'. The goal of authentic balance between the requirements of the indigenous forest communities and the specialised requirements of the larger national community (Sinha, 1987) can not be achieved only through bureaucratic intervention and statistical (paper-based) 'conservation' of forests in the Ruler's interest; real People's initiative and their active participation is the essential precondition in this respect.

It is worthy of mentioning that the concept of social forestry is very old especially in India. Lord Buddha about 2500 years back exhorted his followers to plant at least one tree in their homestead and nurse the same for five years. Prior to this, vedic literatures are replete with terms like 'gramya van' (village forestry), 'Vatikas' (gardens), 'Vihars (sanctuaries)' etc. (Sen and Jha, 1994). But after thousands of years, what are the realities today? Along with the advancement of 'civilisation', rampant deforestation and other anti-forest activities of 'civilised' society pose a grave threat to the existence of (hu) man!

Forests have a tremendous role in Human Development. Physical health, mental health, population structure, family structure, food habit, culture, values, economy, politics etc. of human societies – all are being affected in many ways by the attitude of modern man towards forest. Hence, proper maintenance of forest is also a sine qua non for real Human Development, i.e. the all-dimensional development of human societies. The use of forests for the well-being of the people in the earlier days has turned into the awful exploitation of forests for the profit of the anti-people section of the 'civilised' society today.

Deforestation–Afforestation Dichotomy

Forest is being used by the forest-dwellers for their own daily requirements. Among others, two types of principal usage are: (i) fire-wood (fuel), and (ii) fodder (leaves) for their live-stock. This forest scenario (so-called 'deforestation'!) caused 'anxiety' to the urban-chieftains globally. But, side-by-side, the major deforestation-activities are initiated by these chieftains (Rulers) themselves! "The advance of cultural liberty must be a central aspect of human development, and this requires going beyond social, political and economic opportunities since they do not guarantee cultural liberty" (UNDP, 2004). But this ethical principle has totally been ignored by the 'civilised' rulers. In one hand, perpetual deforestation is going on for the benefit of the Government exchequer and mafia-contractor-dishonest political leader unholy nexus, sacrificing totally the cultural values and human needs of the forest-dependent people. In India, the actual forest-area is significantly less than that of the recorded forest-area (Raha, 2004). It is not in India alone that deforestation is going on. The rapid deforestation is going on in different corners of the globe. It is not only the people in India who are responsible for clearing the forest and making the earth barren, 'civilizations' of different countries are equally responsible for the same. Within a period of ten years (between 1979 and 1989), the rate of deforestation in 34 countries has increased to 90 per cent (Raha, 2004). On the other hand, the quick growing variety of trees are being cultivated, in the name of much brokered 'afforestation', which are not beneficial for the forest-people : neither these can be used as fire-wood or provide food to the forest dwellers, nor the leaves can be used as fodder for the live-stock. Moreover, these quick growing varieties of trees (e.g. Eucalyptus, Ãkãshmoni, Sonãjhuri) require more water; as a result – underground water-level has sunken down, which leads to scarcity of water. This anti-adivasi, anti-people, unwanted so-called 'developmental' cycle is going on at the behest of the 'civilised' Rulers!

'Conservation' of forest through a bureaucracy started since British regime. As a result, the adivasi people became the 'encroacher' in their own traditional home; on the contrary, the aboriginal people considered the rulers/administration as 'intruders' on their natural and inherited rights. ''This gave rise to movements and suppression'' (Deogaonkar, 2004).

Development is inconceivable in Anthropology ''if it is not able to improve the quality of life of a people. Infrastructural development is a meaningless phenomenon when it is devoid of a direct relation with the well-being of the people'' (Piplai, 2004). The success and/or failure of deforestation/afforestation should also be measured from the broader perspective of total Human Development.

Role of Foreign 'Donors'

It is high time to assess the role of foreign 'donor' agencies in the context of deforestation/afforestation phenomena. The 'foreign donor agency' like World Bank, USAID, SIDA, CIDA etc. are investing thousands of crores of rupees in India for the forest-centric activities of different kinds. For example, in the State of Jharkhand, a project of Rs.1,147.77 crores for the forestry sector to be spent over a period of 10 years to be funded by the World Bank has already been sent to the Government of India (Ghosh, 2004). The simple question must be answered is that: is it 'donation', or, it is a kind of 'investment' in the interest of so-called 'donor' foreign agencies/countries? From Nagaland in the east to Gujarat in the west, from Jammu & Kashmir in the north to Tamilnadu in the south–total 6,40,107 sq km of Indian land-mass is covered by six types of forests (Manorama, 2003). The importance of these forests in Anthropology is due to the fact that these forests protect and nourish a large number of people. When both the deforestation and afforestation programmes endangered the life of the people living in and around many of these forests, and served the interest of the 'donor' countries (and the upper strata of the society) in many ways, this pertinent question can not be averted, especially due to the terrible problem of the sons of the forests, i.e. the adivasi people.

From the academic discussion it is clear that in the year 1990, 15 per cent of the forest in India, in 17 Districts of 4 States (Madhya Pradesh, Maharashtra, Orissa & Andhra Pradesh) were "Under Naxalite Control" (Ghosh, 2004). In 2005, it is now being admitted in different media that the forests of 156 Districts of 15 States are presently bearing the 'Naxalite Presence'. It is to be noted that, the 'conservationists' of various shades, with their different ideological frames, are simultaneously and gradually becoming more active along with the growing 'tribal unrest'. The point of importance here is that, many of these 'active conservationists' are getting regular and good amount of monetary help (in many ways) from various foreign 'donors' which, the present author presumes, is intended to dampen this growing 'Naxalite menace' as well as the wrath of the foresters.

Considering this total canvas of socio-political reality, the actual role of the foreign 'donors' should be considered.

Conclusion

All the modern phenomena like global warming, 'peace-keeping' war (e.g. Afghanistan or Iraq), globalisation of the rampant exploitation of forests in the different corner of the globe, etc. is throwing a profound challenge towards Anthropology, which is concerned not with the Ruler's interest but with the life of the People. To meet this challenge is the call of the day.

Indian Forest Act was enacted in 1865. Though this act was subsequently amended but not completely repealed, the basic idea reflected in 1894 forest policy was conservation and protection of forests for state revenue and to restrict the local community's rights and privileges in the forest (Basu, 1987). Participatory researches in this area concludes: "The major conflicts between the tribals and the Government in the recent past are centred round, 'those who command forest land and forest resources'. The situation has created an antagonistic relationship between the local community and the state machinery" (Basu, 1987).

The Anthropologists can play a greater role in this challenging area of human survival, which has hitherto remained neglected and/or is given a very passing importance. In the age of so-called 'globalisation', the time has come to leave the ivory tower and open the eyes in this mundane and virgin field for the development and upliftment of the people in general and the downtrodden forest-dependent people in particular, and to save the forest from wanton destruction which may lead to disaster in the near future.

REFERENCES

Basu, N.G., *Forests And Tribals*, Manisha, Calcutta, 1987, pp. 15, 173, 176.

Deogaonkar, S.G., *Environmental Preservation and Tribal Population; in : The Dying Earth*, ABC Publications, Kolkata, 2004, pp. 473-474.

Ghosh, Abhik, Forest and Watershed Management in Jharkhand; in : *The Dying Earth*, pp. 347, 353.

Monorama, *Monorama Year book - 2003*; Kottayam, p. 579.

Penguin, *The Penguin India Reference Year book - 2005*, New Delhi, p. 723.

Piplai, Dipak, *People's Anthropology of Life : A Conceptual Approach*; Paper presented at the First Indian Anthropological Congress, Shillong, 2004, p.3.

Raha, Manis Kumar, Saving the Dying Earth; in : *The Dying Earth*, pp. 129, 134.

Sen, P.K. and L.K. Jha, Concept of Social Forestry; in : *Forestry for the People*, Ashish Publishing House, New Delhi, 1994, p. 1.

Sinha, Surajit C., Foreward; *Forests and Tribals*, Manisha, Calcutta, 1987, p. ii.

UNDP, *Human Development Report, 2004*; United Nations Development Programme, Oxford University Press, New Delhi, 2004, p. 6.

15

FOREST CONSERVATION STRATEGIES
Economic Development and Human Welfare in Jharkhand

DINESH KUMAR MISHRA

Introduction

The newly created state of Jharkhand is endowed with enormous amount of natural resources including the Forest resources. The vegetation cover of Chhotanagpur plateau is witnessed by highly rich plant diversity which provides a very large platform for sustainable economic development aimed at human welfare. The Saranda forest, which happened to be the largest Saal forest in Asia, has reduced to considerable extent due to anthropogenic activities. Other patches of forest existing in East and West Singhbhum also project a very grim picture with reference to their unabated exploitation and lack of effective policy to substitute them. According to Asheim (Nick Hanley, 1997), sustainable development is a requirement to our generation to manage the resource base such that the average quality of life we ensure ourselves can potentially be shared by all future generations. Every nation of the world has an objective to get rid of poverty, hunger, economic backwardness and lower standard of living of its people. There is no denying the fact that sustainable utilization of natural resources in general and biological/forest resources in particular provides a background for achieving sustainable economic development, which is reflected in human welfare.

Forests of East and West Singhbhum

According to a recent report (Singh – 2002), a comparative view of entire Jharkhand State, and two districts (East & West Singhbhum) can be obtained in terms of total area, total forest area and percentage of forest cover. It shows that West Singhbhum (including Seraikella and Kharsawan) has highest percentage (34.46%) of land area covered by forest whereas the East Singhbhum covers only 29.70 per cent of its land with forest. These two figures are higher than the average forest area of the Jharkhand State. The following table provides the comparative account of three different parameters of Jharkhand State, East Singhbhum and West Singhbhum.

Place	*Total Land area (Sq Km)*	*Total Forest area (Sq km)*	*% of Forest Cover*
Jharkhand	79714	22,046	27.66
East Singhbhum	3533.35	1049.40	29.70
West Singhbhum (+ Seraikella)	9907	3413.98	34.46

It is apparent that the forest area is high in West Singhbhum because of the presence of Saranda Forest and less industrial activities. On the other hand, East Singhbhum registered lesser forest area due to large-scale industrial activities and deforestation for infrastructure development and human settlements.

Plant Diversity in the Forests

The plant diversity in East and West Singhbhum is very rich. Various plants having different economic values are discussed here. Different economic values include timber, fire-wood, medicinal fibre/textile, forage etc. Species of trees, shrubs, herbs and grasses with their respective economic uses are as follows–

Sl. No.	*Name of Plant (Local name)*	*Botanical names of Plants*	*Important Economic Uses*
1	Asan	*Terminalia tomentosa*	Widely used as a timber. Furniture, door frames, Rearing of Silkworm.
2.	Saal	*Shorea robusta*	Used as Timber, Leaves used as plates, Katora, Fruits yield, oil, Gum used as Dhoona, Tender branches used as Datwan.
3.	Sagwan	*Tectona grandis*	Fine Quality Timber.
4.	Amla	*Phyllanthus emblica*	Fruits are rich in vitamin-c, high medicinal value, fruits are prepared as Morabba
5.	Arjun	*Terminalia arjuna*	Timber value. Barks used for curing blood pressure and Heart disease.
6	Ram Datwan	*Smilex macrophylla*	Medicinally used against Rheumatism and pains.
7	Bahera	*Terminalia bellerica*	Fruits have high medicinal value.
8	Bhela	*Semecarpus anacardium*	Black juice is used for making natural dye. It is also used against skin disease.
9	Bel	*Aegel marmelos*	Fruits are edible Gum is also extracted from fruit.
10	Babla	*Acacia arabica*	Gum is extracted` and used for various purposes.
11.	Ber	*Zizyphus jujuba*	The fruit is delicious and many rural people make money out of ripe and unripe fruits.
12	Bija Saal	*Pterocarpus marsupilum*	Timber value. Used as medicine against Diarrhoea. Leaves used as fodder.
13	Palas	*Beutea monosperna*	Gum used in diarrhoea. Flowers are the source of a rich natural colour. Wood sometimes used for house construction.
14	Kendu	*Diospyros melanoxylon*	Fruit is edible and poor people earn money by selling ripe fruits.

Sl. No.	*Name of Plant (Local name)*	*Botanical names of Plants*	*Important Economic Uses*
15	Piyal	*Buchnania latifolia*	Fruit is edible. Very popular among local people. Wood is used as timber.
16	Gular	*Ficus flomerata*	Fruits are produced in large number and eaten by the local people. Leaves are used as fodder.
17	Harre	*Terminalia chebula*	Certain organs are used as medicine. Fruits are used to prepare "Trifala Churna". Gum is used to manufacture tannin material.
18	Jamun	*Syzigium cumini*	Fruits are edible and poor people earn money out of ripe fruits. Seeds used as medicine against diabetes.
19	Kadam	*Anthocephalus cadamba*	Fruits are eaten
20	Kusum	*Schleichera trijuga*	It has timber value. Seeds are eaten. Seeds are used to extract oil.
21	Karanj	*Pongamia glabra*	Pods yield oil (Karanj Tel) It is used for skin disease.
22	Paasu	*Cleistanthus collinus*	Branches are used for construction of House. Used as fuel wood by local people.
23	Chiraita	*Andrographis peniculata*	Whole plant is used as medicine for blood purification and stomach disorder.
24	Van Tulsi	*Ocimum americanum*	Source of perfume and used against skin disease.
25	Neem	*Azadirachta indica*	All parts used as medicine against measles, skin disease, stomach disorder.
26	Mahua	*Madhuca indica*	Flowers used to extract raw wine, wood used as fuel, as well as timber.
27	Ritha	*Sapindus trifoliatus*	Fruits are used as detergent.
28	Sisir	*Sterculia villosa*	This is a fibre yielding plant.

Sl. No.	*Name of Plant (Local name)*	*Botanical names of Plants*	*Important Economic Uses*
29	Simal	*Bombax malabaricum*	The wood is used as a packing material. It is also used for making boat.
30	Shisam	*Dalbergia sissoo*	It has enormous timber value.
31	Baigna	*Vitex nigundo*	Used as medicine for Gout. Leaves are used against insects.
32	Kachnar	*Bauhinia purpurea*	Flowers have ornamental value. Bark is used for making fibre.
33	Sar	*Saccharum arumdinacem*	Solid stems are used for making arrows.
34	Aam	*Mangifera indica*	Fruit is eaten
35	Kans	*Saccharum spontaneum*	Used for making rope.

Reasons for Forest Degradation

There are various reasons for deforestation. First of all, there is non-implementation of forest laws strictly. Many illegal contractors carry out unlawful activities in felling of important trees with mala fide intention. The local tribals are not empowered to prevent such type of activities. Sometimes, they involve themselves in illegal cutting of trees in spite of the fact that forests provide them all types of necessities. Young saplings are sometimes uprooted or felled to sell and earn money. There are people who openly transport the useful tree products by local trains despite knowing that such activities are unlawful. The law-enforcing agencies are not to be seen around. The forest lands are being used for constructing houses and other unauthorised activities. Some of the other reasons of forest degradation is forest fire. Most of the forests of Dalma and Saranda are prone to fire during summer season. From middle of February till the onset of monsoon, a large area of the forest is denuded by wild fire. Sometimes, people with mala fide intention set on the fire, which brings about damage to thousands and thousands of saplings which would make the forest more dense in due course of time. The dry fallen leaves are responsible for quick spreading of the fire in all directions, because they are highly inflammable. This way, the forest fire reduces the forest productivity and new succession of vegetation is set in motion. Moreover, domestic livestock grazing forms a great part of biotic pressure on the forests. In rainy season, grazing incidence affects the forest crop badly by destroying and crushing of new forest crop – shoots and exposing the upper soil for easy erosion by rain water. As such, the lower and tender crops are uprooted during grazing process. Illegal collection of honey, thatching grasses, tubers, wild mushroom, Rhizomes, roots, satmul, bhumikohra, anantmul etc. are also responsible for forest degradation in Jharkhand.

Some Suggestions for Forest Conservation Strategies

1. Any strategy for forest conservation can never be a success story without the involvement of the locals and tribals. Efforts must be made to provide apprenticeship to the unemployed educated young men. Such persons should be trained to identify each plant and learn their various economic uses. They should also learn how to identify the encroachment areas.
2. Effort should be made to establish local forest research institutes through the involvement of experts, botanists and environmentalists. Such institutes will generate database of forest flora and suggest conservation methodologies.
3. People of different socio-economic status should be ensured with permanent livelihood so that they should channelise their energy towards the protection of forest resources.
4. Each forest, including their adjoining areas should be developed as a tourist spot so that economic activities are initiated and employment generation are made. Such initiatives would ensure forest conservation.
5. Creation of eco-development zones is extremely essential for the conservation of forest. According to a report prepared by Department of Forest and Environment, Government of Jharkhand (Deo-2000), the following measures should be taken for the creation of eco-development zones. Efforts should be made to make people aware about forest protection and raise their socio-economic status. The eco

development plans should be circulated to each government departments from block to district level of East and West Singhbhum.

6. The Non-Government Organisations should also be actively involved in the dissemination of eco-development programmes. The district authorities should be given strict instructions for effective implementation of such plans.

Conclusion

The above discussion sharply underlines the fact that economic development through the conservation of forest resources is the need of the hour for the Jharkhand State. Forest Conservation Strategies would also help in the restoration of deteriorating environment. The eco-restoration programmes would ensure favourable climatic conditions. Such desirable conditions will certainly increase the scope of cultivation of other crops for economic development. It is hoped that the present communication would invite the attention of concerned authorities for taking needful action in this direction.

REFERENCES

Bandhu Desh, Garg R.K.: *Social Forestry and Tribal Development,* Indian Environmental Society, Darya Ganj, New Delhi, 1986.

Deo Niranjan Prasad, BFS: *Management Plan of Dalma Wild Life Sanctuary*: Department of Forest and Environment Government of Jharkhand, 2000.

Ghate S. Ruchas: *Forest Policy and Tribal Development*, Concept Publishing Company, New Delhi, 1992.

Malhotra V.K., Bihar Forest Laws, Malhotra Bros, Fraser Road, Patna, 1998.

Mamoria Dr. Chatrubhuj and Jain Dr. S.C.; *Vikash Evam Pryawarn*, Tatha Bhartiya Arthbewastha, Sahitya Bhawan Publications Agra, 2005.

Nick Hanley, J.F. Shogren and Ben White: Environmental Economics, Theory & Practice, Macmillan India Ltd, 1997.

Sharma Atul 1991; *Pryavarn Aur Van Sanrachhan Samasya Evam Samadhan*, Taxila Publisher, New Delhi, 1991.

Sikliger Dr. Punamchand 1994; *Van Evam Adiwashi Samajik Jiwan*, Shiva Publisher Distributors, Udyapur, 1994.

Singh Suresh Kumar: Jharkhand 2002, Readers Corner, Fraser Road, Patna, 2002.

Virotam Dr. B.; *Jharkhand Itihas Evam Sanskriti*, Bihar Hindi Granth Academy, Patna, 2001.

INDEX